Cambridge Studies in Biotechnology

Editors: Sir James Baddiley, N. H. Carey, J. F. Davidson,
I. J. Higgins, W. G. Potter

5 Immobilized cells: principles and applications

Immobilized cells: principles and applications

J. TAMPION
M. D. TAMPION
School of Biotechnology, Polytechnic of Central London

The right of the
University of Cambridge
to print and sell
all manner of books
was granted by
Henry VIII in 1534.
The University has printed
and published continuously
since 1584.

CAMBRIDGE UNIVERSITY PRESS
Cambridge
New York   New Rochelle
Melbourne   Sydney

CAMBRIDGE UNIVERSITY PRESS
Cambridge, New York, Melbourne, Madrid, Cape Town,
Singapore, São Paulo, Delhi, Tokyo, Mexico City

Cambridge University Press
The Edinburgh Building, Cambridge CB2 8RU, UK

Published in the United States of America by Cambridge University Press, New York

www.cambridge.org
Information on this title: www.cambridge.org/9780521292535

First published 1987
First paperback edition 2011

A catalogue record for this publication is available from the British Library

Library of Congress Cataloguing in Publication data
Tampion, John.
Immobilized cells.
(Cambridge studies in biotechnology; 5)
Bibliography
Includes index.
1. Immobilized cells – Industrial applications.
2. Immobilized cells.   1. Tampion, M. D.   11. Title.   111. Series.
TP248.25.I55T36   1988   660'.6   87–13214

ISBN 978-0-521-25556-1 Hardback
ISBN 978-0-521-29253-5 Paperback

Contents

Authors' note

In quoted work, the original authors' scientific names for organisms have been retained.

Introduction

Definition and scope

Although the word 'immobilized' is now part of the language of Biotechnology its scope is delineated more by practical examples than by precise definition. Bucke (1983) focuses attention upon the prevention of free cell movement within the liquid phases of a reactor system. Mattiasson (1983) gives greater emphasis to the matrix in, or upon, which the cells are immobilized, and also to the point that the cells may be at one extreme fully capable of division or at the other so debilitated as to possess only a single type of enzyme activity. For the purposes of the present book an immobilized cell is defined as a cell, or a remnant thereof, that by natural or artificial means is prevented from moving independently of its neighbours to all parts of the aqueous phase of the system under study. Such a definition is sufficiently broad as to include virtually all the commercially significant examples, but would also include many examples of natural immobilization, such as multicellular organisms and communities of microbial organisms. It is therefore necessary to add the further qualification that the purpose of such immobilization must be for use in the production or service industries, i.e. biotechnology (for definitions of which, see Bull, Holt & Lilly, 1982). It is easy to see why authors move rapidly to the details of the subject and avoid such wide-ranging definitions. It is, indeed, the economic exploitation of immobilized cells which is of special significance although there is no doubt that the enormous effort being devoted to them will lead to advances in many other, apparently unrelated, areas of scientific study.

Difficulty is sometimes experienced in distinguishing between conventional fermenters and immobilized cell reactors. A fermenter, regardless of its exact type, is a culture vessel designed to be used firstly to increase the biomass of the active organisms with which it has been inoculated. This growth may accompany or precede the conversion of growth substrate into some useful metabolic product. In the batch mode of operation the fermentation is terminated at a time considered appropriate and the cycle of inoculation, growth and harvesting repeated. Such a regime is satisfactory for many products, particularly those produced late in the growth cycle of the culture. Products of the latter type are often described as being non-growth-related, species-specific or

1

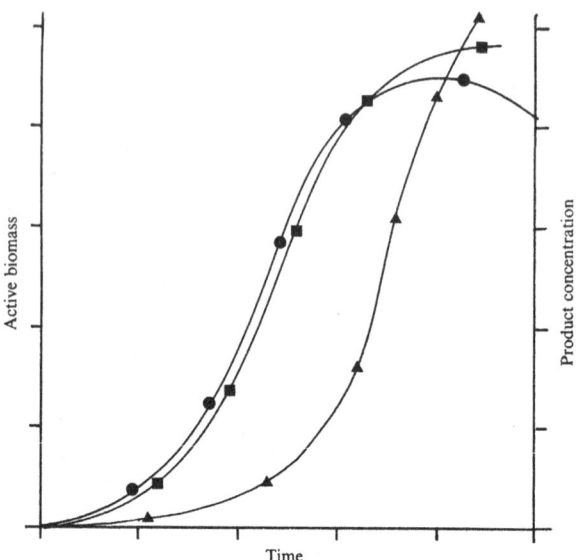

Figure 0.1. The time-course of growth (●) and the production of a typical growth-related (■) and non-growth-related (▲) fermentation product in a batch fermenter.

secondary in nature (Figure 0.1). In the continuous mode of operation growth-related substances or biomass itself can be readily produced but many metabolites of economic interest are often not produced in sufficient quantities. In either batch or continuous mode it may be possible to use the active biomass produced in the fermenter to carry out the biotransformation of a separately added precursor rather than the total *de novo* biosynthesis of a particular molecule. In some cases the cultured organism may not even be capable of the total synthesis. In such a situation it is clear that the living cells are being used simply as a biocatalyst. One of the major problems is, however, that the biocatalyst is generally not very stable and catalytic activity is often lost quite rapidly after the cessation of cell growth. Much of the industrial and academic research into enzyme and cell immobilization has been carried out in the search for 'extended-life biocatalysts'. It has been found by experiment that free cells grown up in a fermenter may have their useful biocatalytic life considerably extended by immobilization (Figure 0.2). The reasons for this increased stability are not entirely clear. Returned in the immobilized state to a vessel essentially similar to the original fermenter the system is an immobilized cell reactor. The boundaries between these uses of the two terms are, however, somewhat blurred. A fermenter may encourage the natural adsorption of living cells onto its internal surfaces,

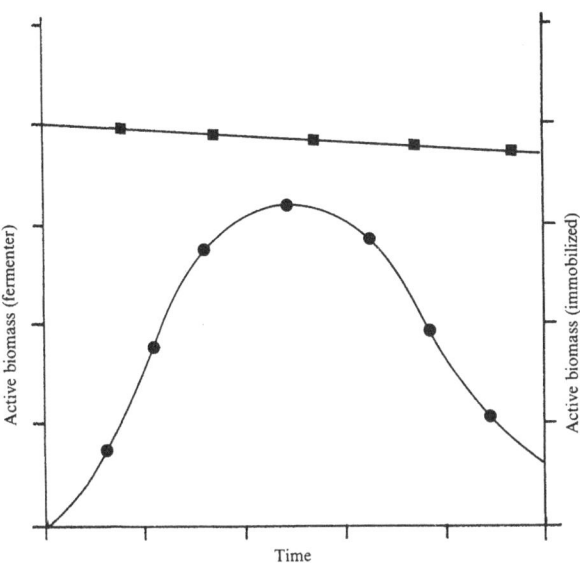

Figure 0.2. Time-course of the change in the amount of biocatalytically active biomass present in a typical batch fermenter (●) and an immobilized cell reactor (■).

or additional support material may be intentionally added, either in sheet or particulate form. Active biomass is then retained if the fermenter is operated continuously. An alternative strategy might be to recycle the active biomass after it has been separated from the liquid culture medium outside the fermenter. The use of flocculent strains of micro-organism may allow the liquid medium to be run off without the cells. Conversely, it is possible to reactivate a debilitated living immobilized cell system by adding fresh culture medium to allow regrowth, *in situ*, of the immobilized cells.

Another viewpoint on immobilization is to regard it as a strategy for process intensification (Atkinson, Black & Pinches, 1980). It is generally accepted that processes using biocatalysts suffer from severe disadvantages compared to simple inorganic catalysis. Chief amongst these are the low concentrations of active biomass and the enormous quantity of water generally present. These lead to low rates of reaction (per unit of reactor volume) and high product recovery costs. Immobilization can pack more active biocatalyst per unit of reactor volume and hence reduce reactor size and cost (Figure 0.3).

The development of techniques for cell immobilization has followed that of enzyme immobilization. In the latter case the enzyme is first separated from cells or cell debris and then, in a crude or purified form,

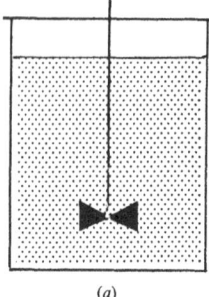
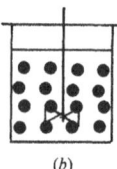

(a)   (b)

Figure 0.3. Diagram illustrating the principle of process intensification by immobilization, resulting in higher volumetric productivity for the same quantity of active biomass. (a) Free cell reactor; (b) immobilized cell reactor.

immobilized. The literature on immobilized enzymes is very extensive and falls outside the scope of the present book. Only in a few special cases, where the more advanced state of knowledge is relevant, will reference be made to work on immobilized cell-free enzymes. The interest in immobilized cells draws attention to the fact that due to economic or scientific reasons there are drawbacks to the use of the comparable immobilized enzyme systems. Those with a knowledge of enzyme immobilization techniques will note that many of the methods can be used also for cell immobilization. This is no more than is to be expected from the nature of the reactive groups involved. In general, however, techniques for cell immobilization are cheaper and easier to carry out. This is because the enzyme separation stage is entirely avoided and the functional activity of the enzymes is protected from damage by a multitude of other substances present, together with the enzyme, in a cell. In addition, many alternative and often physiologically mild techniques are available for cell immobilization.

Cellular material for immobilization

Virtually any cell type or subcellular organelle is a possible candidate for immobilization. The limiting factor in the first instance is likely to be the availability of sufficient cells of the required type. Provided they can be conveniently cultured microbial cells present few problems and many are very readily available and, for that reason alone, have been extensively studied. The ease of supply of conventional brewer's or baker's yeast (*Saccharomyces cerevisiae*) has made it one of the most common organisms for the purely academic investigator. Industrialists have, naturally enough, chosen to work on organisms of direct relevance to some already existing or proposed processes. Certain service industries, such as wastewater treatment, have always based their biological processes upon mixed cultures of micro-organisms which show a natural immobilization characteristic. The natural flocculation of yeasts has been of fundamental significance to the production of potable alcohol for centuries. Many micro-organisms attach themselves to the walls of culture vessels and several types of fermenters have been specifically designed to encourage this tendency, allowing simple biomass retention and the development of continuous processes so that the fermenter used for initial growth becomes an immobilized cell reactor by mere change of operational procedure.

As the artificial techniques for cell immobilization developed attention turned to the cells of animals and plants, where the problems of availability are much greater. Techniques for the culture of plant cells are well advanced and suitably selected cell lines of many species are already grown in fermenters. The preparation of large quantities of plant cells is, however, by no means easy. Because of the slow rate of cell division and the need for exceptional care in maintaining sterility, progress has not been as rapid as with microbial immobilization, but the possible rewards of success have interested many academic and industrial laboratories (Brodelius & Mosbach, 1982). One of the major considerations, however, is the extent to which the plant cells grown in culture are able to carry out the metabolism of the differentiated cells typical of the original plant (Fuller & Bartlett, 1985). It may well be that the process of immobilization, simulating more closely the conditions existing in an intact plant, can help solve this difficult problem. Plant cells, by their possession of a cell wall, are comparable to microbial cells and similar techniques of immobilization may be used. Due to the absence of a cell wall, animal cells are inherently more fragile. The problem of the bulk culture of many normal animal cells is further aggravated by the possession of two attributes: the need for a support surface to grow on and contact inhibition. Some typical forms of cell are given in Figure 0.4. When a normal animal cell in culture contacts another cell its further division is inhibited, leading to low cell yields from culture. The development of microcarrier systems (see, for example, Pharmacia, 1981; Hirtenstein & Clark, 1983) has allowed natural immobilization to be exploited as a means of obtaining high yields. Abnormal cells, such as those in cancerous cell lines, may be grown more readily in bulk but are incapable of producing many of the important substances of the anchorage-dependent cell lines. Few of those working with immobilized cells have the necessary skills to cope with the difficulties of animal cell culture.

When the cell wall is removed from a plant or microbial cell the protoplast, surrounded by the plasma membrane, is left as a functional unit. This is comparable to an animal cell in terms of its response to

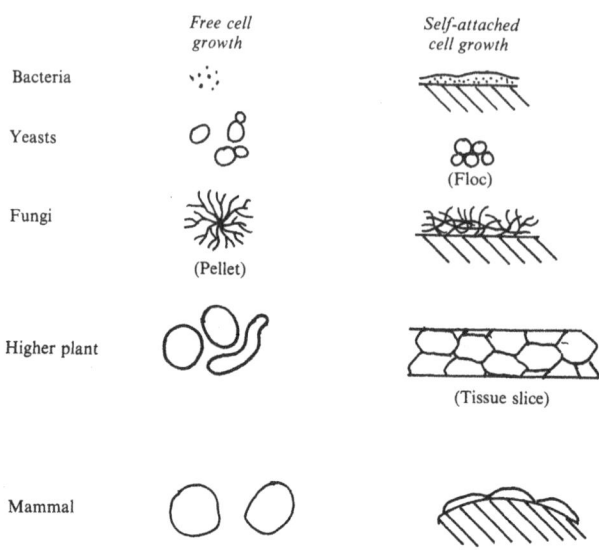

Figure 0.4. Some typical physical forms of cells used in immobilization studies.

osmotic shock and has occasionally been the subject of specialized immobilization studies. The continued integrity of the plasma membrane of any cell is essential if it is to remain capable of growth and division. It is possible to alter the selective permeability of the plasma membrane to allow the freer diffusion of substrates into the cell or metabolites out of it. This process is described as permeabilization and may be achieved by a wide variety of different techniques. With mild treatments only a relatively slight change in the selective permeability of the cells will occur. This will generally be sufficient to prevent further cell division but still allows the majority of the cell's metabolism to continue and may even stimulate the activity of certain metabolic pathways. With harsh treatments the metabolic integrity of the cell will be destroyed and only certain facets of enzymatic activity will remain. These remnants are, however, often considerably stabilized and even activated, by the treatment. The process of permeabilization may be carried out before, during or after immobilization.

Quite distinct from the process of permeabilization is the prior disruption of living cells by an appropriate technique of cellular homogenization and the recovery of separated organelles, or fragments thereof. Conventional techniques for purification of the fractions can be applied and the resulting preparation immobilized. Much of this work is carried out for purely academic investigations, albeit of great scientific

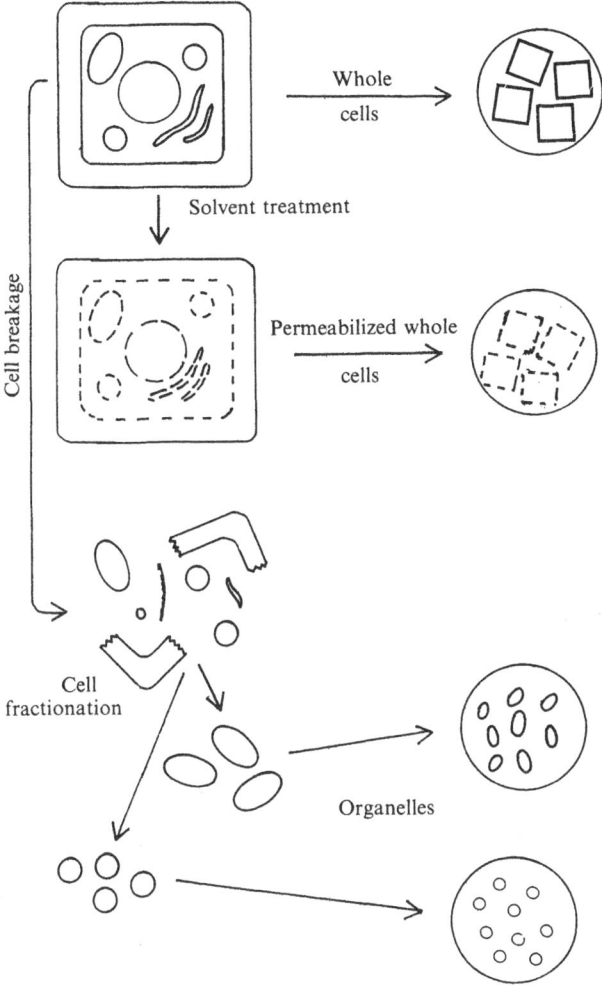

Figure 0.5. Some possible approaches to the immobilization of the biocatalytic activity of cells by entrapment.

merit, with industrial application a long way in the future. Chloroplasts, mitochondria, microsomes, peroxisomes, chromatophores and other membrane fragments have all been investigated (see, for example, the review article by Tanaka & Fukui, 1983). Various combinations of immobilized enzymes, subcellular fragments, organelles and intact cells have also been tried, often with components taken from different species (Figure 0.5).

Immobilization methods

A simplified classification of immobilization methods is given in Table
0.1. Adsorption of cells onto a solid surface is probably the mildest of cell
immobilization techniques. In its simplest form it is also one of the
cheapest methods and therefore of particular use where an industrial
process does not result in a 'high added-value' product. The success of the
technique depends, in the first instance, upon the properties of the cells
themselves. The natural evolution of species has produced many
organisms that are capable of adhering to surfaces. Some of these cause
problems to man, as in the microbial fouling within pipes and on all
manner of other surfaces exposed to an aqueous environment. An
important method of wastewater treatment, the trickling filter system,
has made use of this property for decades. In adsorption there is generally
an initial weak attachment of the cells, which can be easily reversed. This
is followed by the development of stronger (multiple attachment)
binding. Often extracellular material produced by the cells is important in
fixing the cells to the adsorption substrate and this may be followed by a
natural entrapment of the cells in a biopolymer matrix. In the older
industrial processes the selection of the appropriate support material has
been largely fortuitous and has relied upon the organism's own ability to
attach to the surface. The more recent active interest in the adsorption
process has led to the study and development of all manner of support
materials, ranging from simple crude mineral substances to complex
ion-exchange derivatives of organic polymers. In addition, the physical
form of the support has received considerable attention, particularly with
regard to porosity and the shape of the material. Coating support
materials with other substances or mixing the cells with various
polyelectrolytes (often natural polymers) can increase cell adsorption.
One of the interesting features of adsorption is that it is not only microbial
cells which adhere. Cultured cells from higher plants and animals also
show this property, often to a remarkable degree. Many of the newer
supports for adsorption are essentially non-selective. Indeed, this very
fact makes them useful for a wide range of applications. An alternative
strategy, however, is to make the support extremely selective. This may
be considered as analogous to the process of affinity chromatography,
now widely used by biochemists. As far as the cells are concerned it is very
mild. Only those with the appropriate surface reactive groups will be
bound, but the binding can be strong under operational conditions.

It is a simple observation in the preparation of alcoholic beverages that
yeasts are capable of flocculating. In this process, the individual cells are,
in effect, adsorbed onto one another. A similar process occurs during the
settling of sewage sludge. The process has been extensively studied in the
brewing industry and, by strain selection, it is possible to control the time

Table 0.1. *A simple classification of cell immobilization techniques*

Technique	Some typical Advantages	Disadvantages
Adsorption		
Neutral supports	Cheap	Cell leakage
	Mild	Sensitive to pH changes
	Reusable	
	Simple	
Charged supports	Mild	
	Reusable	
	Simple	
Flocculation	Simple	Cell leakage
	Mild	Diffusional limitations
Entrapment		
Natural polymers	Mild	Diffusional limitations
	Simple	
Synthetic polymers	May be simple	Toxicity
		Expensive
		Diffusional limitations
Covalent coupling	Permanent	Toxicity
		Expensive
Containment	Mild	Diffusional limitations
	Simple	Expensive
	Reusable	

at which this occurs and so influence the properties of the product. Under continuous culture conditions, by means of relatively simple design features of the vessel, it is possible to select flocculating strains of organisms which are not generally considered to show this phenomenon. A further extension is to add polyelectrolytes or other suitably active material, and initiate flocculation artificially. Again there is a long-standing and extensive literature on this topic from the traditional brewing and wastewater treatment industries. The chief interest in the present context is its application to the development of new industrial processes. Although fundamentally different in its origin it is convenient to mention at this point the process of pelletization. It has been found by experiments that certain filamentous micro-organisms can be encouraged to grow as discrete, spherical mycelial masses (pellets) rather than as a single mat. This is usually achieved by using a spore inoculum and carefully controlling the cultural conditions. In particular, a relatively strong agitation and the absence of 'dead-spots' in the liquid flow pattern may be favourable. This cannot really be considered as an immobilization technique but may lend itself to process intensification and the development of processes based on extended biocatalytic activity.

Gel entrapment, because of its mildness, ease of operation and wide applicability, has been one of the most studied of all immobilization techniques. As the name implies, it is merely the trapping of cells within a three-dimensional matrix such that the pores in the matrix are smaller than the cells. It follows, therefore, that the matrix (or lattice) must be constructed *de novo* around the cells rather than the cells adsorbed into a preformed porous material. A wide range of materials is available for entrapping cells, ranging from natural polysaccharides and proteins to purely synthetic polymers such as polyacrylamide. In general, it may be said that the cells are trapped within a gel and that the gel may be permanent or reversible. Soft and reversible gels may have poor functional properties while the permanent ones may involve potentially toxic stages in their preparation. In some techniques the gels are further stabilized after preparation. Obviously, where full cell viability needs to be retained the milder methods must be used, but for many potential applications it may be adequate to retain only a very limited amount of enzymic diversity in the cells. Where full viability is retained it will be possible to increase the active biomass within the gel by supplying appropriate nutrients and growth conditions, thus promoting process intensification. Excessive growth in a weak gel may, however, disrupt the matrix and allow cell leakage.

One problem encountered in using cells entrapped in a gel is the diffusional resistance of the gel to substrate and products. A logical means of reducing this is to have only a thin layer of material around groups of cells, which are free to move within their own particular capsule. The advantages of larger particle size are retained by this encapsulation, but still with some diffusional problems. Two major categories may be distinguished. In the first are those where the retaining capsule barrier separates two essentially similar liquid phases. The barrier may be of a lipophilic material, simulating the type of membrane which already surrounds a cell, but with much simpler permeability properties. Alternatively the capsule could, perhaps, be of synthetic or natural polymers. A second approach uses liquid two-phase systems, made by separately preparing two solutions of different polymers, such as a dextran and a PEG (polyethylene glycol). When mixed they are able to form an emulsion of one aqueous phase within another. By adjusting the properties of the two phases very interesting partition effects may be achieved so that products, which might cause inhibition of the catalytic activity, may be effectively removed from the reaction environment into the other phase.

An extension of the membrane separation principle is to have two compartments to the system. There are a number of different methods of achieving this. Obviously, the surface area of the membrane, in relation to the mass of active biocatalyst present, and the length of the average diffusion path from one compartment to another, are important. The

Table 0.2. *Classification of support materials for immobilization*

Type	Category	Typical examples
Inorganic	Natural	Alumina, magnesia, silica, zirconia
	Manufactured	Activated carbon, brick, ceramics, coke, glass
Organic	Polysaccharides	Agar, agarose, alginate, carrageenan, cellulose, dextran, pectate, wood (with lignin, etc.)
	Proteins	Collagen, egg white, gelatin
	Synthetic polymers	Polyacrylamide, phenolic resins, polystyrene, polyurethane

systems can be considered as an extension of the dialysis techniques used in conventional biochemistry. The greater efficiency afforded by using a large number of fine hollow fibres, in place of a single large chamber, has led to a number of studies with this technique. The equipment is rather expensive and requires skill in its operation but has great potential. Animal cells, for example, have been cultured in hollow-fibre systems and this may lead to their use as immobilized cell reactors for high-value products (Hopkinson, 1983). Microbiological applications are also being explored.

The technique of covalent coupling – the creation of permanent chemical bonds – has been extensively used in enzyme immobilization. When applied to cells it has the disadvantage of being a generally harsh technique which results in loss of viability of the cells. For many applications, however, this may be of no consequence. A wide spectrum of possible techniques is available. As one example, a support material with appropriate mechanical and physical properties can be used (Figure 0.6). The term 'grafted' is used to indicate that such a support has been coated with a reactive coupling agent. Used in the 'ungrafted' state the same support may well be capable of directly immobilizing cells by adsorption. Various types of non-biocatalytic material, such as cheap soluble proteins, may be mixed with cells, in the absence of a particulate support material, and covalent bonds formed *inter alia*. The resulting immobilized cell preparations may be cast in the form of membranes or as particles of an appropriate shape. It is even possible to cross-link covalently just the cells to one another, providing greater stability to the aggregates than can be achieved by flocculation.

The previous paragraphs have briefly considered the various methods of immobilization according to the process involved. It will be clear that certain supports (or carriers, as they are sometimes called) may play a role in more than one method. These supports may be classified into organic and inorganic materials and each category may be further subdivided according to the details of its composition (Table 0.2). This sensible

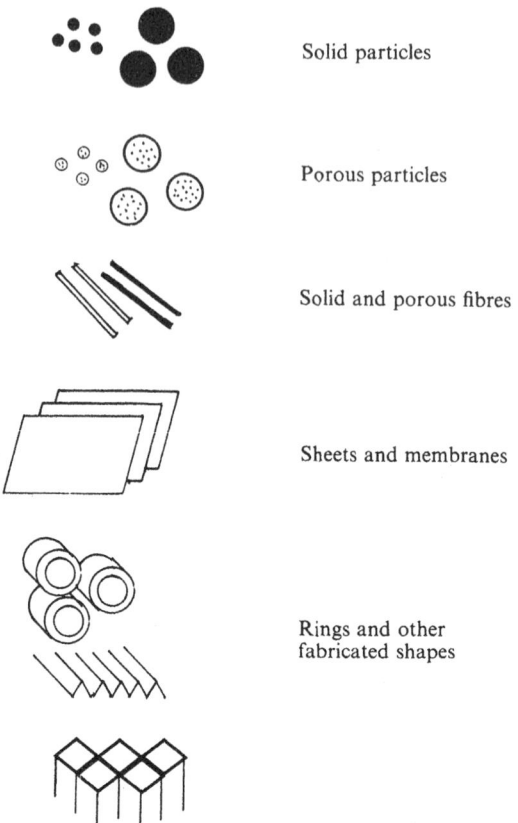

Solid particles

Porous particles

Solid and porous fibres

Sheets and membranes

Rings and other
fabricated shapes

Figure 0.6. Some typical physical forms available for immobilized cell preparations.

approach to assist in the rational selection of support materials has been discussed by Kolot (1981).

Characteristics and advantages of immobilized cells

Having discussed the immobilization process, by definition and example, it is now appropriate to consider briefly the features required for a successful application of the technology and the general advantages to be obtained from cell immobilization. Two features are of paramount importance for the economic exploitation of immobilized cells. Firstly, the preparation must possess high and relatively long-term biocatalytic activity. Except where regrowth of the immobilized cells is possible, there is no method of increasing the actual amount of biocatalyst present after immobilization. Even where regrowth is possible it is clearly something of a disadvantage to have to stop the biocatalytic process to regenerate active biomass. Attention therefore focuses upon the biocatalytic activity after immobilization compared to that before immobilization. This is described in more detail in Chapter 2. After immobilization the activity of the preparation will inevitably decrease because proteins, which are the active biocatalytic molecules, are not completely stable. It is not convenient to determine the time taken for complete loss of biocatalytic activity because the decay curve is non-linear. By analogy to the process of radioactive decay it is usual to determine the 'half-life' of the immobilized preparation – the time taken for 50% of the original biocatalytic activity of the immobilized cell preparation to be lost. Depending on the system and conditions this may vary from hours to years. Rather than quoting direct chronological time it may be more realistic for industrial users to quote the number of process cycles (often involving more than one operating condition) over which the preparation is stable.

The second feature of importance is the durability of the preparation. Although this will influence any measurements of biocatalytic activity it can reasonably be considered as quite distinct and relates to the functional characteristics of the immobilized preparation. If a tall reactor with static contents, such as a packed-bed, is being used the particles will compress under the weight of the material above. This may seriously affect flow rates and hence reactor efficiency. With a fluidized-bed or stirred-tank reactor the movement of particles over one another and the vessel walls will result in abrasion. Cells will be removed from their supports and supports themselves worn away. Both of these, as well as inadequate or reversible immobilization, may also result in contamination of the product stream. This is most undesirable. Resistance to abrasion but without excessive brittleness, is therefore desirable. In some cases there is the additional problem of the production of gas bubbles by the immobilized cells. These may create regions where the substrate is excluded from the biocatalyst or even lead to the physical rupturing of the immobilized cell preparation. As noted previously, however, many immobilized cells are so debilitated as to have lost the capacity for gas production.

Another aspect of durability is resistance to microbial degradation. Because the biocatalyst is inherently susceptible to biodegradation a suitable operational strategy must be used. The cheapest solution is to operate under conditions where microbial contaminants cannot easily develop. This may be achieved by the use of elevated temperatures, extremes of acidity or alkalinity, or very high substrate concentrations. The use of soluble biocides is in general not convenient due to the possibilities of product stream contamination or adverse effects on the biocatalyst. In an exceptional case the natural antimicrobial activity of a support material (egg white) has been considered particularly

Table 0.3. *Generally desirable features for immobilized cell preparations*

High biocatalytic activity
Long-term stability of biocatalyst
Possibility of regenerating biocatalyst
Low loss of activity during immobilization
Low leakage of cells
Non-compressible particles
High resistance to abrasion
Resistance to microbial degradation
Low diffusional limitation
Spherical shape (most applications)
High surface area
Appropriate density for the reactor type
Simple techniques are involved
Cheap support materials
Non-toxic materials are used

advantageous. Because of the additional costs and precautions necessary it is only under unavoidable circumstances that fully sterile process lines are used. Prior sterilization of the substrate stream may prove an adequate precaution in some cases.

Assuming highly active and durable preparations, the major restrictions on bioconversion frequently centre around the problems of the diffusion of substrates to the site of biocatalysis and of the products away from the site. The length and resistance of the diffusion pathways have been topics of intense interest, particularly to those with appropriate mathematical backgrounds. It is impossible to obtain a system which is optimal for all criteria. Generally, durability is economically more significant than complete minimization of diffusional restrictions (Table 0.3).

For a successful commercial process some additional features are of interest. In general, both the immobilization process and the overall production system should be simple and capable of appropriate scale-up. Low-volume, high-value products are clearly able to sustain more complex systems than high-volume, low-value ones. A factor often overlooked by the academic researcher is the safety of the industrial operators of a full-scale process, who may be less technically proficient than a research scientist. In any case, there are considerable differences between handling a few millilitres and several hundred or even thousands of litres. Often, the industrialist will prefer substances already accepted for use in food or food manufacturing, even if the proposed product is not for consumption. Problems with chronic toxicity or allergenicity are then much less likely. Above all, there must be a reasonable chance of the process proving economic. Many good ideas and excellent support materials have been produced which are totally uneconomic for large-scale use. These points should be borne in mind in the later pages of this book. Many of the results referred to are from academic laboratories where the criterion of success may be the production of results suitable for publication in a reputable scientific journal rather than the potential economics of an industrial scale-up. This is, of course, how it should be. Without the comparative basic operational facts there is little possibility of rational choice by the industrialist. Many aspects of immobilized cells remain to be elucidated.

It is virtually impossible to produce accurate economic comparisons between processes involving immobilized cells and the various possible alternatives. So much depends upon the costing system adopted and uncontrollable market forces. Certain generalizations are, however, possible regarding the relative advantages and difficulties. The technology of batch fermentation is well understood and the equipment readily available. Although immobilized cell processes can be operated on a batch mode it is generally more desirable and economic to use them as continuous processes. As noted previously, one of the major advantages of immobilization is the extension of the life-span of the metabolically mature, stationary phase cells. Because continued cell growth is often not required the medium in which immobilized cells are kept can be relatively simple, making product recovery easier. The wasteful cycles of medium sterilization, inoculation, cell growth and equipment cleaning found with a batch fermentation, can be significantly reduced. Due to process intensification the size of the vessel can also often be greatly reduced for a comparable immobilized cell process. In comparison to a continuous fermentation the reduction in vessel size may be less dramatic but the benefits gained from the saving in media components and the ability to operate under non-growth conditions are considerable (Table 0.4). The biosynthesis and biotransformation of substances not readily achieved by continuous fermentation may be easier with immobilized cells. The major problem likely to be encountered with the immobilized cells is the growth of contaminating micro-organisms. In many cases this difficulty may be alleviated by the ability of the immobilized cells to function under more extreme conditions than the comparable free cells.

The fact that cell immobilization has developed after enzyme immobilization highlights the economic benefits. As mentioned earlier, there is a complete saving on the cost of enzyme separation and purification. Frequently, the enzymes present in immobilized cell preparations may exhibit greater stability than the same enzymes immobilized in a pure state. One difficulty, however, of the cell process arises from the need to have a larger reactor, due to the presence of non-catalytic biomass. Even worse may be the contamination of the

Table 0.4. *General comparison of immobilized cell processes with conventional batch fermentation and immobilized purified enzyme processes*

Processes	Features
Immobilized cells	Long-term stability
	Often cheap to prepare
	Usually reusable
	Good for multi-enzyme processes
	Some ability to regenerate biocatalyst
	Good for complex products
	Less familiar technology and industrial equipment but smaller size
	Generally medium capital costs for equipment
	Cheap product recovery
Immobilized purified enzymes	Long-term stability
	Expensive to prepare
	Difficult for multi-enzyme processes
	Cannot be regenerated
	Usually reusable
	Not suitable for complex conversions
	Less familiar technology and industrial equipment
	Generally medium capital costs for equipment
	Cheap product recovery
Conventional batch fermentation	Unstable system
	Often cheap
	Good for multi-enzyme processes
	Complete new fermentation needed
	Good for complex products
	Uses familiar technology and industrial equipment
	Generally high capital costs for equipment
	Expensive product recovery

product stream with unwanted cell debris or unwanted biotransformation products resulting from the presence of a multitude of enzymes.

For certain products there are alternative chemical processess. In general, these require higher temperatures and pressures and, except for the very simplest molecules, are likely to be less specific, leading to the production of unwanted by-products. Conversely, the immobilized cell processes may have much higher costs for product recovery, due to the generally lower concentrations of products and the higher water content. There are some examples of the use of cells in non-aqueous media but the presence of liquid water is essential for most biocatalytic processes. To compare an immobilized cell process with a conventional agricultural or

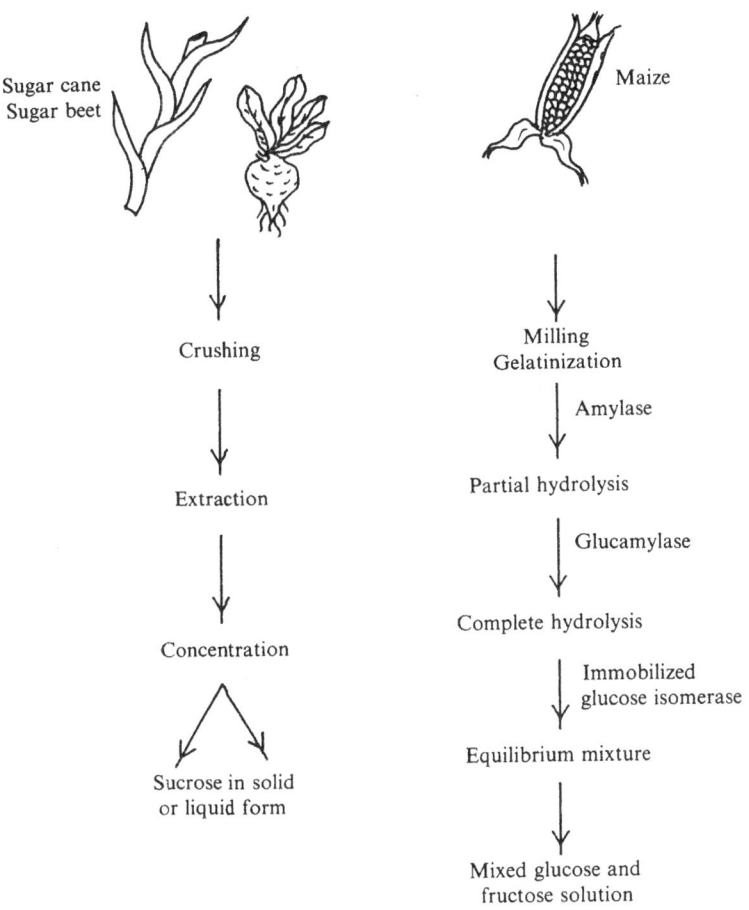

Figure 0.7. Comparison of the stages in the High Fructose Corn Syrup process with those in conventional sucrose production.

horticultural alternative is barely possible. The structure and economics of the two systems are so different and the anomalous influence of political factors so great that true market forces do not operate. The classical example of this is the production of High Fructose Corn Syrup using immobilized cells in comparison to sucrose production from sugar cane or beet. In countries where free market forces operate the former process is used extensively. In others, by means of specific legislation, the agricultural system is supreme (Figure 0.7). It is, however, true to say that an immobilized cell process does have the advantages of being independent of seasonal and climatic influences, because it is carried out inside buildings, throughout the year. This greater stability and continuity of supply may be of great significance in relation to products that are otherwise obtained from regions of the world showing political, economic or other instabilities. This may more than offset the greater investment in equipment and skilled manpower required for any immobilized cell process. Several important medical applications of immobilized cells are worth noting. They may enable the economic production of a wide range of active compounds used in the treatment of diseases and functional disorders. They may also be directly useful as a replacement cell therapy or as part of an external detoxification system. A suitable method of immobilization may prevent the activation of the immunological defence system and rejection of the cells, although it is too early to report success on this front. The development of biosensors incorporating immobilized cells coupled, by a variety of means, to sophisticated electronic instrumentation is currently receiving much attention. Some may have medical applications, others for pollution monitoring or as probes in fermentation vessels.

The scientific study of immobilized cells is a topic of intense interest throughout the world. Studies of the immobilization process itself and the operational stability of the resultant biocatalysts abound (see, for example, Chibata & Wingard, 1983). They have drawn attention to the lack of knowledge of the influence upon metabolism of restricting free cell movement. Cellular differentiation may also be affected and the functional alterations caused by immobilization or permeabilization methods also pose equally interesting problems. Much of the immobilization research of the past has been essentially empirical but there is currently a developing interest in these more fundamental aspects, which will undoubtedly extend present knowledge of cellular physiology.

1   Cell biology of immobilized cells

Any consideration of cell immobilization has two primary facets: the cell and the immobilization system. The cell is the basic unit of life and may be described as a complex molecular aggregation separated from the external environment by a selectively permeable membrane. The essence of life resides in the ability of this molecular complex to reproduce itself, at the same time transmitting its information content to the daughter cells. This ability to divide is the ultimate and true test of cell viability. It depends on the possession, by the cell, of a totally intact metabolic capability. To describe a cell as viable, however, does not mean to say that it is actually dividing, nor even that it ever will divide under the operational conditions of a given immobilized cell reactor. It implies only that it is potentially capable of division. There are two fundamentally distinct types of cell that are distinguished from one another by a whole series of differences. They are named, however, according to the primary distinction of the way in which their heritable information content is organized. The simpler of the two types are the prokaryotes, which includes the bacteria and the cyanobacteria (formerly called the cyanophyta or blue-green algae). These do not possess a distinct nucleus, separated from the rest of the cell contents by a nuclear membrane. They never possess the cell organelles known as mitochondria and plastids. The majority of organisms belong to the other category, the eukaryotes. With the exception of a few highly differentiated cell types within certain multicellular organisms the cells of the eukaryotes have a distinct nucleus. They also contain mitochondria and, if from the Plant Kingdom, one or more of the various types of organelles belonging to the plastid group. It is generally considered that mitochondria represent the end-point of an initially symbiotic relationship between a eukaryotic pro-cell and a bacterium. There are many reasons for believing this to be so and these are now well-documented in the literature dealing with the early stages of the evolution of life forms. For essentially similar reasons, the plastids including the chloroplasts are thought to derive from originally symbiotic cyanobacteria.

An alternative way of categorizing cells is to distinguish those that possess a cell wall surrounding the plasma membrane from those in which the plasma membrane is directly exposed. The former are commonly described as 'plant' cells (including the prokaryotes) and the latter as

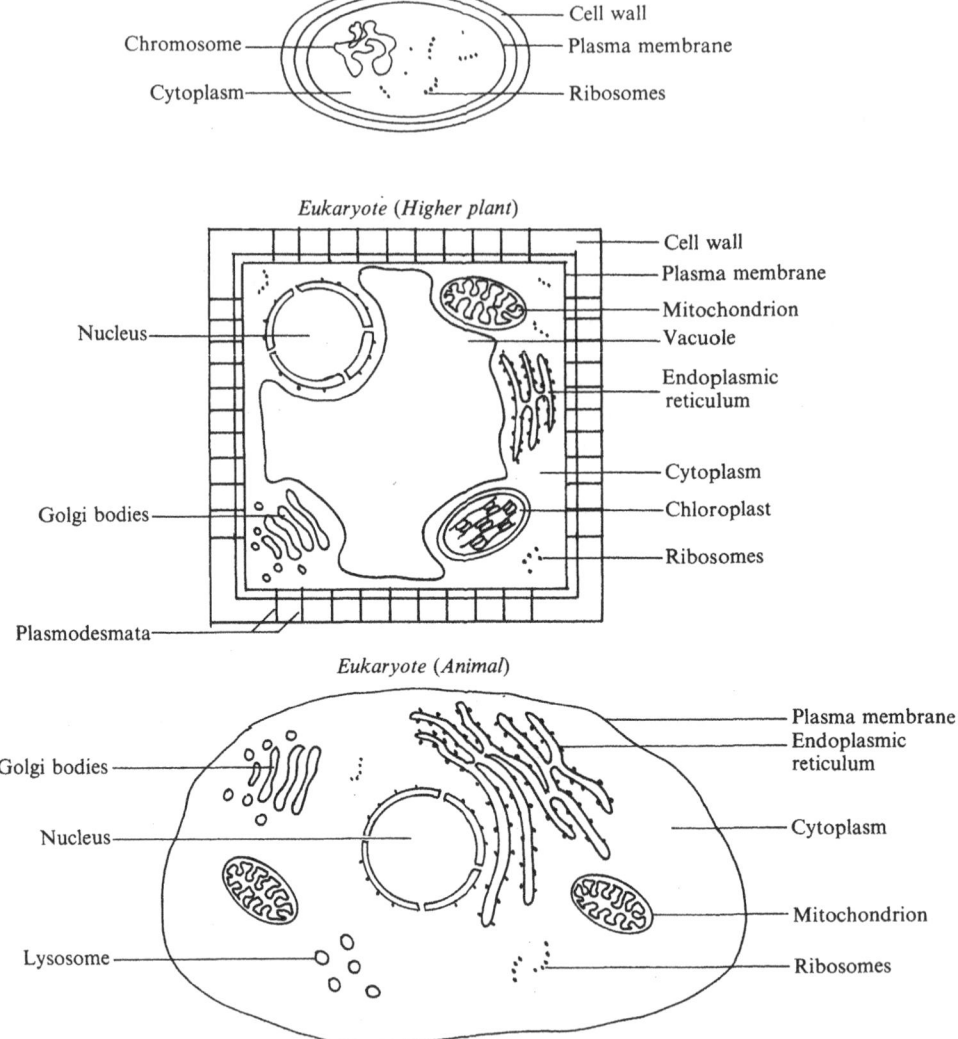

Figure 1.1. Diagrammatic representation of the ultrastructure of generalized prokaryotic and eukaryotic cells.

animal cells (Figure 1.1). Certain mutant forms of plant cells have lost the ability to produce an intact cell wall. An important function of the cell wall is to limit the osmotic swelling and prevent bursting of the cells. Without a cell wall this occurs when cells are placed in a solution of lower osmotic strength than the cell contents. There are now well-established

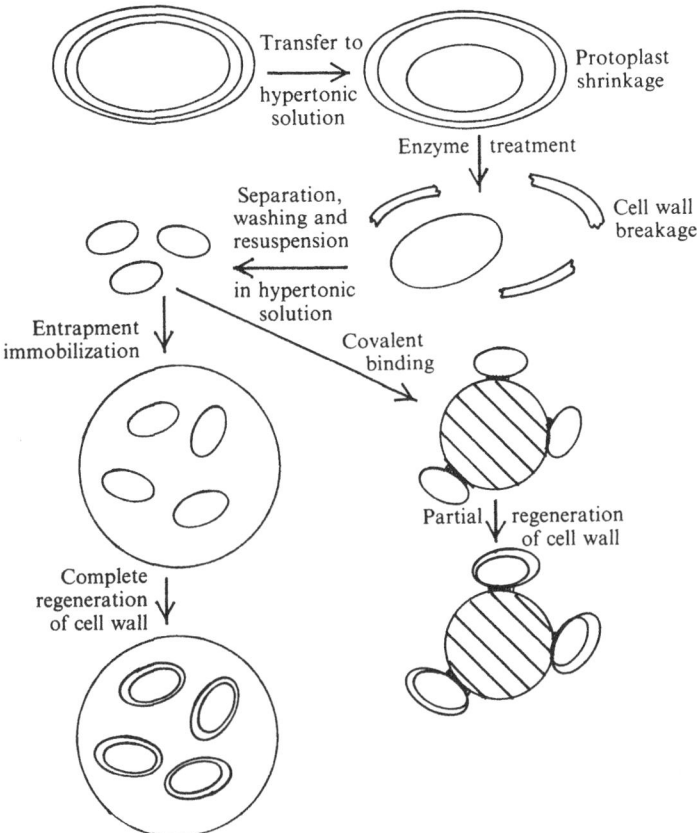

Figure 1.2. Outline of the main stages in the preparation of immobilized protoplasts.

and fully documented ways of removing cell walls in such a gentle manner that the plasma membrane remains intact and the cell is still viable. Such cells, which have been artificially denuded of their cell walls, are called protoplasts. Given time and appropriate conditions, a new cell wall will be synthesized by the protoplast and the cell will thereby regain its ability to exist undamaged in media of low osmotic strength. Protoplasts, like cells, can be immobilized. Depending on the immobilization method used they will regenerate a complete cell wall or, where certain parts of the plasma membrane have been blocked by the formation of covalent bonds during immobilization, a partial cell wall, which is more permeable than the original complete cell wall (Figure 1.2).

An additional alternative way of grouping eukaryotic organisms is into those with a multicellular construction and those where the functional organism consists only of a single cell. This distinction may seem very simple and easy to determine and, in many cases, it is. In others, however, the distinction is not so clear. In the slime moulds, for example, cells may exist as free units at one stage of the life cycle and come together in a co-operatively acting multicellular form in another. Some colonial unicellular organisms live together in close single-species aggregates which can gain functional advantage from the close proximity of the individual cells. Some may show natural self-attachment to support materials or even self-immobilize themselves in a matrix material of their own extracellular secretions. This is the simplest type of immobilization requiring only the provision of supporting surfaces to achieve immobilization. In the natural situation there are often several distinct species present, either in balanced, relatively constant, proportions or showing cyclical changes in the size of the populations of the respective component species. Some interesting comments on these and related topics are presented in Quayle & Bull (1982).

Even the cells of the simplest single-celled organisms are not necessarily always in the same state. As a cell passes from its first production as a daughter cell to the state of itself being in the act of cell division it shows important changes in its metabolic activities and physiological responses. The term 'cell cycle' is used to describe the changes which occur between one cell division and the next. By convention the cell cycle is divided into four or five phases, designated by a letter code and separable by defined events. The first phase of a daughter cell's independent existence is the first growth phase (G_1) during which time it increases in size and may also show some changes in its metabolic capabilities. This phase is terminated by the onset of replication of the DNA, which occurs during the S phase. The specific enzymes involved in DNA synthesis are active in this phase. When DNA replication is complete the cell enters a second period of growth (G_2). This is, in turn, terminated by the commencement of cell division (CD). In eukaryotic organisms, having a distinct nucleus, this is the process of mitosis (M) and may be divided into two distinct phases. The first involves the disappearance of the nuclear membrane and the visual reappearance of the chromosomes, their partial splitting into two chromatids each and then completely to produce the chromosome sets of the daughter nuclei. This is called karyokinesis (K). The second involves the formation of two distinct daughter cells by the separation of the protoplasm into two parts, each surrounded by a plasma membrane. This is called cytokinesis (C) (Figure 1.3). Generally, cytokinesis follows immediately after karyokinesis.

If karyokinesis occurs without cytokinesis the process of endomitosis is said to have occurred. If the new nuclear membrane encloses the entire, doubled, chromosome complement of the cell this is now an autotetraploid cell. Tetraploid cells may or may not show distinct

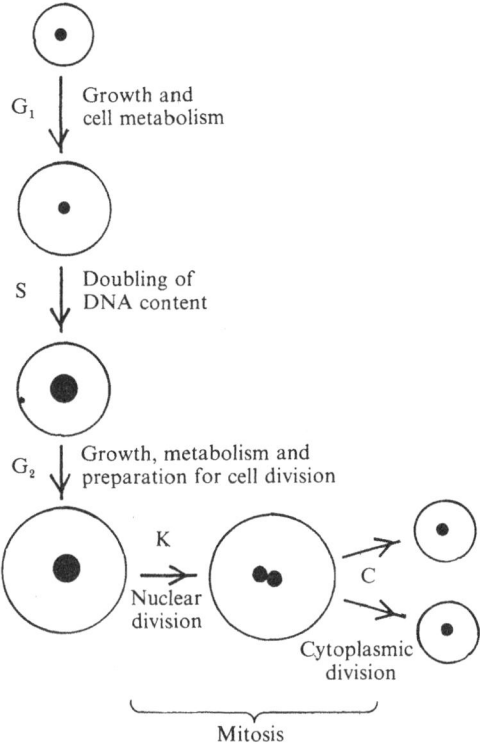

Figure 1.3. Stages of the cell cycle in a eukaryotic cell.

differences in growth or physiology compared to the original diploid cell. Complete autotetraploid plants are generally sterile because of the abnormal meiosis which occurs, due to the possession of four of each homologous chromosome, but they are sometimes of commercial value and are therefore propagated vegetatively. A whole range of cells containing various chromosome numbers may arise if the reformed nuclear membrane does not enclose all the chromosomes or some abnormality in the separation of the chromosomes occurs. Such cells may be capable of growth and even undergo cell division but even if the original organism can be regenerated, it will be expected to be sterile. Vegetative propagation may be used to maintain plants with these abnormal chromosome complements. In general, animal cells are much more sensitive to deviations from the normal chromosome complement and multicellular animals even more so, showing considerable abnormality of metabolism, growth and development. It should be noted that the majority of eukaryotic plants show an alternation of generations. In one the chromosome number is half that of the other. One or the other is usually the dominant generation but in some algae both may be of equal size. The term karyotype is used to describe the full chromosome complement of a cell or organism. In practice karyotyping is generally carried out by photographing cells during the metaphase stage of karyokinesis, cutting up the photograph and rearranging the chromosomes in a regular series. It is to be expected that cells of the same organism, but with differing karyotypes, will probably behave differently in the immobilized state but this has not been investigated.

The cell cycle is of great interest to those involved in culturing isolated plant and animal cells and single-celled micro-organisms. The length of the various phases of the cell cycle vary greatly depending on the particular organism being studied. Organisms with rapid doubling times obviously have shorter phases somewhere in their cell cycles. Partly, of course, this will be related to cell size so that small prokaryotic cells naturally do not need to synthesize as much cellular material before division as do larger cells. Experiments have shown that the major differences occur in the length of the G_1 phase. In the case of certain animal cells, such as those of hamster ovary cell lines, it is known from scanning electron microscope (SEM) and other studies that the surface membrane of the cell changes markedly during the passage of the cell through the various phases of the cell cycle. This can alter the ease with which they can be attached, or detached, from supports. In the case of cells which are surrounded by a cell wall such changes would not, even if they actually do occur, be visible with the scanning electron microscope.

It appears that there is a positive stimulus which causes a cell to pass from the G_1 to the S phase and that once the S phase has commenced the cell is, other conditions being suitable, committed to move forward to complete the cell cycle and division into two daughter cells. Various external conditions, exerting an effect inside the cells, may hold up the progress of the cell cycle. Immobilization itself or the change of conditions, such as the composition of the medium in which the immobilized cells are held, may have this effect. Provided the cells remain viable it is possible to reactivate the progress of the cell cycle. This is often done with cells which have been immobilized by one of the milder techniques. When the biocatalytic activity of the immobilized cell preparation has dropped below an acceptable level it is often possible to 'reactivate' it by adding a culture medium which will support full cell growth, for an appropriate period of time.

An important difference between cells in free suspension culture and the same cells when immobilized, is the extent of cell to cell contact which occurs. With some animal cells it has been shown that cells which stop dividing when they contact with surrounding cells form physical junctions between their membranes, which allow small ions and molecules to pass directly from one cell to another. Immobilization may assist this and so, perhaps, allow cells to become non-dividing and develop metabolic specialization in the G_1 phase. The situation with plant cells is particularly interesting. In the intact higher plant all cells are connected together by plasmodesmata, which provide a direct interconnection between the cytoplasm of adjacent cells. In suspension culture the aim is generally to produce a rapidly growing culture of single cells, without plasmodesmata. Immobilization of such cells will allow the plasmodesmata, which form between daughter cells during cell division, to remain intact. This may be an essential requirement for metabolic differentiation and so be of importance in the development of industrial processes for the production of plant secondary products. Before leaving this brief discussion of the cell cycle it is important to note that ordinary cultures of isolated cells or micro-organisms contain cells in all the various phases of the cell cycle. This is because the phases do not always last for exactly the same length of time and the process of division does not produce two absolutely identical daughter cells. There may be as much as a ten per cent difference in size between the two daughter cells (Mazia, 1974). It is possible to obtain cells all in the same phase by special treatments. Such cultures are called synchronous cultures and are the only convenient experimental material for the study of the cell cycle.

It is now helpful to describe briefly the ultrastructure of 'typical' prokaryote and eukaryote cells and the functional significance of the various structures within them. For a fuller discussion the reader is referred to any modern textbook of cell biology, although it should be noted that many of these may be deficient in certain important aspects, usually those concerning the cell biology of higher plants. A good account of the major intracellular organelles of plant cells is given in Hall, Flowers & Roberts (1981).

Prokaryotes

Considering firstly a typical bacterial cell, this will contain a single chromosome consisting of a circular double strand of DNA, with sufficient bases to code for a few thousand different protein molecules. When supplied with appropriate nutrients the cell cycle may be completed in as little as 20–30 minutes. The actual division of the cells is preceded by the attachment of the two chromosomes formed during the S phase to the cell membrane at opposite ends of the cell. In many bacteria the individual cells remain close to one another after division and may form 'colonial' filaments or aggregates. Sometimes these may be held together by mucilaginous secretions from the cells.

In the actinomycetes, containing such commercially important genera as *Streptomyces*, the process of cell division does not go to completion and an elongated filamentous structure is produced containing many double strands of DNA distributed along its length. Because of their greater size it is physically easier to immobilize such 'hyphal' strands and it has even been advocated that the normal separation of bacterial cells might, with advantage, be prevented by treatment with appropriate toxic substances (such as Crown – see Tso & Fung, 1981) that produce a filamentous form of growth. Such a treatment is not, of course, a permanent conversion of one growth form into another. Even the actinomycetes may produce separate short cells under appropriate conditions. The other major structures visible within a heterotrophic prokaryotic cell, using the electron microscope, are the ribosomes. Each ribosome consists of a large 50S and a smaller 30S subunit, with the S in this case referring to Svedberg units, determined by the sedimentation behaviour of the particles in an ultracentrifuge. The ribosomes normally occur in groups called polysomes and are the site of translation of the messenger RNA into protein structure.

There are no structures within prokaryotic cells which are equivalent to mitochondria or plastids. This, of course, is part of the basis of the theory that these two types of organelles found in eukaryotic cells represent an advanced state of an originally symbiotic relationship between, respectively, bacteria and cyanobacteria and a 'host' eukaryotic cell. The cell membrane of aerobic bacteria is the functional site of its aerobic respiratory electron transport chain, in addition to the normal functional role as a selectively permeable membrane. In photosynthetic bacteria and certain other species with very specialized metabolic capabilities there may also be internal, membrane-bounded vesicles and infoldings of the cell membrane Some bacteria are motile, but this is obviously a feature that is incompatible with the immobilized state. Movement may be simply physically prevented. Movement systems such as flagella and cilia may also be inactivated by toxic substances or functionally lost by the selection of mutants.

One of the most commonly used staining procedures for bacteria, the Gram stain, relies on differences in the composition of the cell wall material which lies outside the cell membrane. If this were simply a matter of the presence or absence of a reactive group it would, perhaps, be of little significance but it relates also to many other important differences. In general, Gram-negative bacteria are more resistant than Gram-positive bacteria to the damaging actions of certain antibiotics and toxic chemicals. The Gram-positive cell wall is the simplest, consisting of a thick layer of the material known as mucopeptide, mixed with techoic acids and other polysaccharides. This material is also called peptidoglycan by some authors. Very little protein is found in the cell wall. *Bacillus subtilis* is a typical example of a bacterium with this type of cell-wall structure. In contrast the thinner cell wall of Gram-negative bacteria consists of an outer membrane of phospholipid and protein composition surrounding a thin peptidoglycan layer. Some connections

Gram-positive bacterium

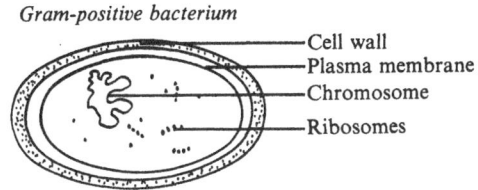

— Cell wall
— Plasma membrane
— Chromosome
— Ribosomes

Gram-negative bacterium

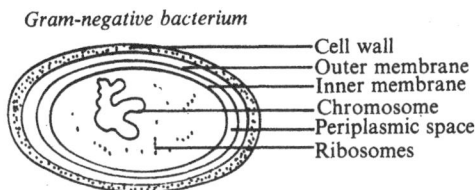

— Cell wall
— Outer membrane
— Inner membrane
— Chromosome
— Periplasmic space
— Ribosomes

Figure 1.4. Diagrammatic representation of the ultrastructure of typical Gram-positive and Gram-negative bacteria.

between the outer membrane and the cell membrane have been observed (Figure 1.4). For a fuller discussion of the outer membrane the reader is referred to Inouye (1980). Protoplasts of bacteria can often be obtained, in isotonic media, by treatment with the enzyme lysozyme, which breaks bonds in the peptidoglycan layer.

The relevance of this distinction in the structure of the wall is that while Gram-positive bacteria excrete a wide range of hydrolytic enzymes directly into the medium surrounding them, Gram-negative bacteria can store such enzymes in the periplasmic space between the cell and outer membranes. Over 25% of the cell mass of a Gram-negative bacterium may be in the periplasmic space, including also specific binding (transport) proteins for sugars and amino acids. The outer membrane provides selective and non-selective channels for certain ions and other nutrients which are essential for growth, allowing them to diffuse passively into the periplasmic space. It is not, however, a non-selective membrane and therefore may be a significant cause of diffusion limitation in immobilized preparations of Gram-negative bacteria. At the same time, it may protect the cell membrane and substances present in the periplasmic space from damage during the immobilization process. Taken overall, however, it might be expected that the Gram-positive wall structure is more likely to confer protection from damage to the functional capabilities of the cell during immobilization. It should be noted that certain detergents which can solubilize the inner cell membrane leave the outer membrane relatively intact, which may be of some relevance during deliberate permeabilization of immobilized cells. Among the major proteins found in the outer membrane are the porins, so called because they allow small molecules, such as sucrose, to pass readily through the membrane but exclude oligosaccharides with molecular weights of around 900 and over. A considerable amount of information has been obtained concerning the genes which are involved in the production of outer membrane components. Mutants with altered permeability properties are readily obtained and might be of benefit in immobilized cell preparations.

The cell wall is not the only material to be excreted to the outside of the cell membrane by prokaryotes. In many species a further two categories of substances are recognized. Firstly, there are the exopolysaccharides, which vary enormously in composition, depending both on the species and even, in some cases, the growth conditions. Where the attachment of the exopolysaccharide is evidently quite firmly to the cell wall this is normally called a capsule and can be revealed by appropriate staining techniques. The term 'glycocalyx' is used in a rather similar sense to that of 'capsule'. Where the attachment is loose, the exopolysaccharide is more easily removed and its outer edge is barely, if at all, discrete from that of other neighbouring cells. This is spoken of as slime. In addition to these two types of cell-associated polysaccharides there is also a group of substances known as extracellular polysaccharides which diffuse out into the general growth environment. Many of the extracellular polysaccharides are of interest as industrially important products, such as the xanthan gums produced by species of *Xanthomonas*. Exopolysaccharides are known to be important in the self-immobilization of bacteria in surface films and microcolonies (see, for example, Costerton & Cheng, 1982). Only recently has it become possible to examine and study satisfactorily the true extent and nature of these exopolysaccharides.

Experiments have revealed, as discussed in Aaronson (1981) and Gerson & Zajic (1979), that the surface charge of prokaryotes is normally negative. This has been ascribed to the presence of techoic acids, acidic polypeptides and various related substances on the surface of Gram-positive bacteria. The acidic lipopolysaccharides in the outer membranes of Gram-negative and blue-green bacteria, together with acidic polypeptides, contribute to their surface charges. Among the simplest and most generalized techniques for immobilization, as described by Kennedy (1979) adsorption onto ion-exchange resins and hydrous metal oxides both rely upon the surface charge of cells for their success. Krug & Daugulis (1983), for example, used a range of ion exchange resins to immobilize *Zymomonas mobilis* and showed that a macroreticular cationic resin was the most efficient. Interestingly, they found the growth form changed from single cells to multiple cell and filamentous forms. Eventually this growth was so extensive as to reduce the flow through their column bioreactor by bridging between the spherical resin beads. It is appropriate to mention here an unusual approach to overcoming problems of excessive growth. Khachatourians, Brosseau & Child (1982) used a mutant strain of *Escherichia coli*, showing abnormalities of cell division, to produce anucleate minicells which retained metabolic activity but were incapable of growth. An agar gel was used for immobilization and the target enzyme was thymidine phosphorylase. One of the most selective immobilization techniques, affinity bonding, relies on the precise biochemical composition of the outer surface of the cells. All of the above immobilization methods are gentle, non-damaging techniques of immobilization. It is clear that harsher techniques, involving some form of physical or chemical stress, will alter the permeability properties of cell walls and membranes, as discussed in Aaronson (1981).

Yeasts

The ultrastructure of eukaryotic micro-organisms is considerably more complex than that of the prokaryotes. Specific morphological features of the organisms, often accompanied by cellular differentiation, are predominantly used in their identification, in contrast to the tests of metabolic capability which are so important for the prokaryotes. Consideration will firstly be given to the yeasts, which have been extensively studied because of their commercial significance (see, for example, Rose & Harrison, 1969).

Although the term yeast is often used loosely, to mean *Saccharomyces cerevisiae*, there are actually many different genera of yeasts and several hundred species. The feature which they all have in common is that the normal form of the organism is unicellular. Many are clearly related to the fungi, particularly where, as in *S. cerevisiae*, they produce spores in specialized structures which have evident similarity to those of filamentous fungi. The unicellular habit of yeasts gives them a high surface area to volume ratio, which probably correlates with their often extremely high rate of metabolism. Because of differences between the genera the following brief account is particularly biased towards the common brewing yeast (*S. cerevisiae*). For greater details of yeast structure and metabolism the reader is directed to the many excellent treatises available concerned with yeasts. The most obvious internal organelle is a membrane-bounded vesicle, of variable size, called the vacuole. It has several important functions, amongst which are its ability to act as a storage site for polymetaphosphate granules and for lipid-like materials, plus a possible role in the transfer of cellular material into daughter cells during budding. The plasma (cell) membrane encloses a

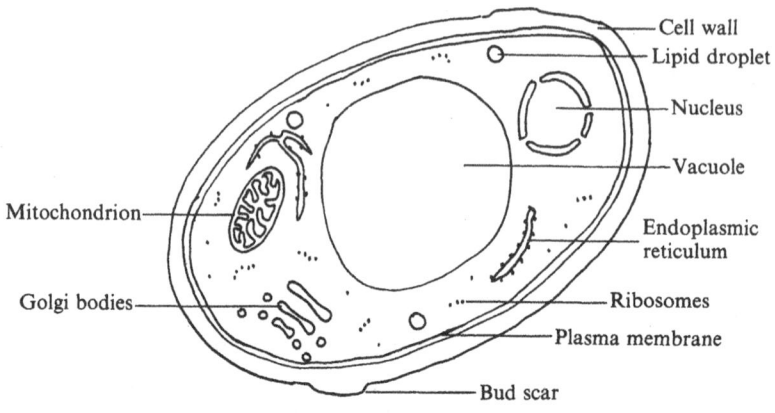

Figure 1.5. Diagrammatic representation of the ultrastructure of a typical yeast.

cytoplasm containing a true nucleus, with a distinct nuclear membrane. The mitochondria, organelles concerned with the process of aerobic respiration, are a prominent feature of the ultrastructure of the cell (Figure 1.5). It is, perhaps, worth noting that mutant forms of yeast, known as 'petite', lack fully functional mitochondria and are therefore incapable of aerobic respiration. In consequence their growth rates are severely restricted. Unlike the situation with anucleate cells of the bacterium *E. coli*, however, they do not appear to have been considered of particular value for immobilization.

There are also extensive membrane systems present in the cytoplasm, associated with the production and/or excretion of high molecular weight metabolites and enzymes. The endoplasmic reticulum (ER) is a double membrane system, generally carrying ribosomes on its surfaces (rough ER). It has a significant role in budding, where the lumen of the ER, or vesicles derived from it, contain important enzymes. It becomes the ascospore plasma membrane during spore formation. At certain stages of the cell cycle distinct Golgi-type vesicles are produced from the ER and fuse with the plasma membrane discharging their contents into developing cell walls. The nuclear membrane is also a double system, continuous in places with the ER system. Distinct membrane-bounded vesicles called sphaerosomes (or lipid granules) are also present in the cytoplasm, particularly in cells at the stationary phase of growth. Lipid material stored in sphaerosomes may be mobilized after their fusion with the vacuole. Other vesicles may also occur. To the outside of the plasma membrane is a substantial cell wall, consisting of two main components. The inner of these is a β-linked fibrillar glucan and the outer is predominantly a matrix-like glycoprotein which is traditionally known as yeast mannan. There may be a small amount of chitin present in the wall.

A variety of enzymes is found in the plasma membrane, others occur in the periplasmic space and the cell wall itself. The surface bears a predominantly negative charge.

A particular feature of certain yeasts is their ability to flocculate, thus bringing about self-immobilization. This varies from strain to strain of a yeast species and also with the state of the culture, being more pronounced in the later stages of a batch culture. Flocculation is generally considered to be brought about by the formation of bridges between divalent calcium cations and insoluble anionic cell wall polymers, possibly through ionic bonding to carboxyl groups. Some yeasts can grow either as discrete cells, which bud off daughter cells or in a filamentous, multinucleate form, sometimes described as 'germ tube' formation, which may be preceded by flocculation of the cells. Flocculation is often used as a simple means of separating spent yeast from the fermentation broth. Weeks, Munro & Spedding (1983), for example, described an improved method for achieving rapid settling of yeast by adding metallic nickel powder and changing the pH value of the medium from acidic to alkaline. The interest of this procedure lies in being able to operate with highly dispersed yeast cells during the metabolically active stage of a process but to be able to recover the cells easily by transient flocculation. This avoids the problems of diffusion limitation found with yeasts which are in a continuously flocculated state. Yeasts can, of course, also be immobilized by any of the methods described in this book.

Many yeasts have a life cycle which is characterized by a true sexual reproductive stage. The normal vegetative cells of *S. cerevisiae*, for example, are diploid. This means to say that they have two of each chromosome in their nucleus. Cultural conditions can cause meiosis to occur and haploid ascospores are then produced. These normally consist of half of the + mating type and half of the − mating type. Fusion of + and − types occurs, normally before release from the ascus, to give the vegetative diploid form again. Under experimental conditions the haploid spores can be prevented from fusing and will give rise to vegetatively budding haploid yeasts, of a specific mating type, used in breeding experiments. It should be noted that this view of the chromosome complement and nuclear behaviour of *S. cerevisiae* is idealized. Some commercial strains have three (triploid) or four (tetraploid) sets of chromosomes or even some incomplete sets. Generally these will not be capable of undergoing meiosis in a regular manner, because of uneven separation of the chromosomes. The karyotype of yeasts is a subject of much debate. Yeasts are normally distributed commercially and to researchers in the vegetative state and any pre-existing chromosomal abnormalities are retained in the particular strain by vegetative (mitotic) division. The process of nuclear divison in yeasts is also unusual, in that the nuclear membrane does not disappear during nuclear division, unlike the situation generally found in eukaryote cells where the nuclear membrane breaks up into small fragments, indistinguishable from small ER segments, during nuclear division.

Fungi

The fungi form an amazingly diverse category of organisms. A few, such as *Mucor* and *Rhizopus* species, can grow in a yeast-like form of discrete rounded cells, although this does not make them, taxonomically, yeasts. The predominant growth form is as hyphae, which may be generally non-septate in the evolutionarily more simple genera, but is septate (but not truly cellular) in other genera.

Close examination usually shows the septa to have a central pore through which the cytoplasm of adjacent units is in contact, and the plasma membrane extended through the pores from one to another. Cytoplasmic material and organelles can pass through these pores. The production of septa is not directly related to nuclear division in ordinary vegetative hyphae but may occur shortly after it. It seems likely that septa are important in the differentiation of fungal mycelia. If the pores are open then maximum tip growth is achieved but plugging of the pores by intracellular material allows those isolated older segments of the hyphae to direct their metabolic capacity towards other ends. Mutants are known (for example, in *Aspergillus nidulans*) which produce defective septa under certain conditions. Nuclear division continues but hyphae grow short and fat rather than extending in the normal manner. Other benefits, such as walling off damaged sections of hyphae, may also arise when pores are plugged and certain fungi appear to have ready-formed structures in their cytoplasm for just this purpose. A wide range of different spore types may be produced, depending on the genera and species being considered and the environmental conditions. Considerable complexity of the life cycle may occur, especially in the basidiomycetes, with several different spore types being produced. Sometimes there is a distinct alternation of the diploid and a separate haploid generation formed after haploid spores are produced. There may also be a separate, quite distinct, dikaryotic phase in which two haploid nuclei are present in one segment of the mycelium. As in yeasts, the nuclear membrane is not lost during nuclear division. In general, only fungi with the simpler type of life cycle are used for immobilization purposes, or if the life cycle is more complex then only one stage of it is used. Many commercially interesting fungi belong to the Fungi Imperfecti, which do not produce sexual spores.

The organelles present in fungi are essentially the same as those of any other heterotrophic eukaryotic organism. Internal membrane systems abound (Figure 1.6), and may have discrete functional roles as

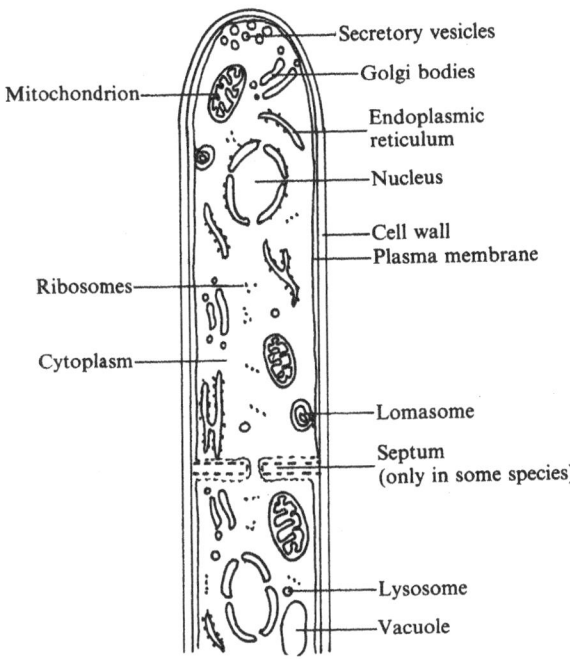

Figure 1.6. Diagrammatic representation of the ultrastructure of part of a typical fungal hypha.

endoplasmic reticulum, Golgi-type vesicles, sphaerosomes, peroxisomes (part of the microbody complex of organelles mentioned earlier in connection with yeast) and lysosomes. The latter are membrane-bounded organelles containing hydrolytic enzymes and involved in the breakdown of particulate material taken into the hyphae by reverse pinocytosis or arising directly inside the hyphae. Because hyphae grow by tip extension there is an age and structural sequence present in hyphae. The vacuolar system becomes obvious only in the older parts, where it may occupy the major part, by volume, of the hyphal segment. Ribosomes, mitochondria and frequently membranous infoldings of the plasma membrane are also present. Outside the plasma membrane fungi possess a definite cell wall and in some cases additional extracellular polysaccharides in the form of a slime around the hyphae. The surface charge, as with most organisms, is negative. The composition of the cell wall varies from species to species and the wall may have many layers. In a typical fungus there may be a layer of chitin present on the inside. Chitin is made up of *N*-acetyl-D-glucosamine subunits and generally occurs in a microfibrillar configuration. Proteins, glycoproteins and various glucans and complex polysaccharides are also present. In a very few cases cellulose appears to

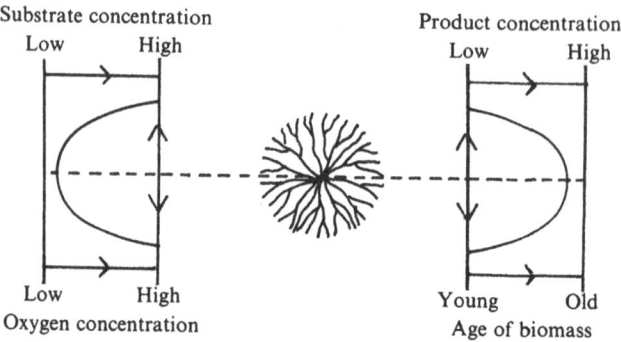

Figure 1.7. Diagrammatic representation of various important gradients in a mycelial pellet.

replace chitin. The preparation of fungal protoplasts is now a well-established technique. Aspects of the cell biology of filamentous fungi are well reviewed in Smith & Berry (1978).

Because of the branching and interweaving of the filamentous hyphae fungi naturally exist as self-immobilized mats covering the surface of any solid growth support. From a biocatalytic viewpoint, however, this is a very inefficient form of growth because there is severe nutrient depletion and diffusion limitation in such mats. Old parts of hyphae, in any case, are generally senescent. In stirred fermenters any surface which is not subjected to sufficient shear forces will become overgrown with mycelium. To overcome this problem, in some species, it is possible to encourage the mycelium to grow in the form of free-floating, spherical pellets. Normally the starting inoculum consists of spores and these are germinated and grown under fairly severe shaking or stirring conditions to prevent the pellets tangling up with one another. The major problem with this approach is the need to keep pellets to a small size, again to prevent diffusion limitation of metabolism and senescence at the centre of the pellets (Figure 1.7). An alternative approach has been to design a special disc fermenter to encourage surface growth of a particular geometry. It should be noted that aerobic conditions are necessary for most fungal fermentations of industrial importance. To overcome the problems of variable hydraulic flow the immobilization of fungi into spherical particles of various materials has been carried out. Broad, Foulkes & Dunnill (1984), for example, allowed spores of *Aspergillus ochraceus* to self-absorb into celite particles and used the preparation to carry out the 11-α-hydroxylation of progesterone. Other researchers have chopped up vegetative hyphae, allowing the cut ends to reseal naturally, and then carried out conventional immobilization of the fragments. In all these cases the reactor conditions need to be such as to limit hyphal growth outside the immobilized particles, otherwise bridging of the particles occurs and the reactor becomes clogged and inefficient. Although mutants affecting hyphal growth are well-known (see, for example, Gooday, 1978) they do not appear to have been used in immobilization studies.

Higher plants

Although there are many thousands of eukaryotic plant species with an evolutionary status between the fungi and the higher (seed) plants, there has been negligible interest in the immobilization of their cells. This arises for a number of reasons but mostly because of a lack of economically-interesting substances in the current range of commercially available products from such species. The situation with higher plants is completely different, however, and many products would find a ready market if they could be obtained from immobilized cell preparations. Even in 1983, Lindsey & Yeoman were still able to describe immobilized higher plant cells as 'novel experimental systems'. To understand the great discrepancy between the commercial exploitation of microbial immobilized systems and similar higher plant systems it is necessary to consider briefly the cell biology of higher plants. If a so-called 'typical' higher plant cell is considered it is seen to contain essentially the same organelles as a segment of a septate fungal hypha: a distinct nucleus, mitochondria, ribosomes, various internal membrane systems and a distinct cell wall outside the plasma membrane. The most obvious difference lies in the possession of an entirely new category of organelle, found in all the eukaryotic photosynthetic plants, from the algae to the flowering plants. These organelles are generally called plastids and several distinct types are known. They all have in common the possession of two concentric membrane systems around their periphery (Figure 1.8).

In cells capable of carrying out photosynthesis the predominant plastids are the chloroplasts. In the vascular plants these always have an ovoid shape. The light-absorbing photosynthetic pigments are present in an internal system of membrane-bounded sacs called thylakoids. The presence of strands of DNA and prokaryote-type ribosomes support the view that plastids represent an advanced stage of symbiosis between blue-green bacteria and a eukaryotic host cell. After periods of active photosynthesis small starch granules are present in higher plant chloroplasts. There is a complex relationship between the various types of plastids. The smallest and simplest type are the proplastids, which have, at the very most, only a few small pieces of internal membrane system. Proplastids are the main type of plastid found in actively dividing cells and are themselves capable of division by fission, ensuring that daughter cells always contain some plastids. Their function is not clearly defined, other

Type   *Major function*

Proplastid   Precursor of other types

Chloroplast   Photosynthesis

Chromoplast   Contains pigments to attract
                                                 other organisms

Amyloplast   Stores and metabolizes
                                                 starch

Etioplast   Partially formed chloroplast
                                                 in dark-grown higher plants

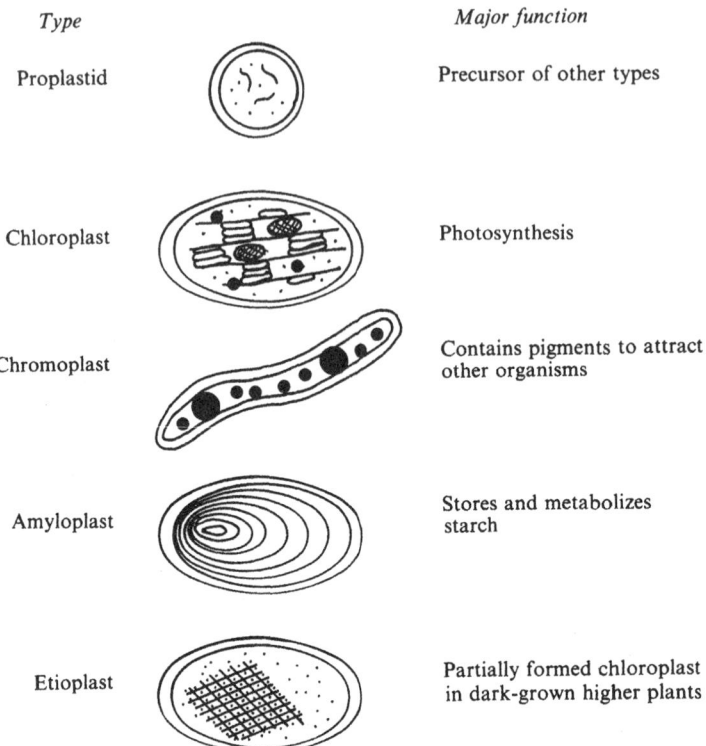

Figure 1.8. Diagrammatic representation of the major types of plastids found in higher plants and their major role in metabolism.

than as being precursors of all the other types of plastid. It should be noted that plastids (and mitochondria) rely upon both their own internal DNA and the cell's nuclear DNA for the production of the complete organelle. Thus, although they can be separated from the rest of the cell and even immobilized, they are not capable of dividing and increasing in numbers outside the cell. In the absence of light the proplastids of flowering plants are not capable of developing fully into chloroplasts. They stop just short of the final stage of differentiation and are then described as etioplasts. These are essentially the same size as chloroplasts but the internal membrane system is arranged in a regular pattern forming the 'crystalline centre', although it is not, of course, actually crystalline in structure. Even a very small amount of light causes the membrane system to be rearranged into the thylakoid system and promotes the final stage of conversion of protochlorophyll into the green chlorophyll pigments.

Chromoplasts are a very variable group of plastids which contain lipophilic pigments belonging structurally to the carotenoids. They are normally filled with carotenoid droplets and in some species protein fibrils also occur. Their major role in the plants is in relation to other organisms because they provide visual attraction for animals, to assist in pollination and/or seed dispersal. Amyloplasts are starch-containing plastids. Starch is the major energy reserve material in most species of high plants and is also implemented in the response mechanism to gravity. In certain storage organs the starch grains (large amyloplasts filled with a crystalline form of starch) may only break down when the cells themselves also autolyse to release nutrients used to support regrowth, or germination of seedlings. In ordinary growing tissues the amyloplasts are themselves capable of both synthesizing starch and breaking it down in a controlled manner, using phosphorylase enzymes which largely retain the energy of the glycosidic bond in a sugar–phosphate bond, unlike the amylases produced during seed germination in endospermic seeds, which are simple hydrolytic enzymes. The various types of plastid are unevenly distributed in higher plants, depending on the function of the particular tissue and organ in which they are found. Under some conditions they may interconvert from one form to another but certain conversions, for example from chloroplasts to chromoplasts during fruit ripening, are more likely than others. In cells in culture, chloroplast development is often inhibited and undifferentiated plant cells in culture are generally heterotrophic, relying on the supply of carbohydrates and other nutrients from the medium. This is exactly the situation of many cells in an ordinary higher plant, where many cells are functionally heterotrophic, even though the plant, taken as a whole, is autotrophic.

Apart from the plastids the major visible organellar difference between higher plant and fungal cells is in the truly cellular nature of the former, together with associated cell wall differences. The septa which separate 'cells' from one another in fungi generally have a large central pore and in only a few cases are cross walls with multiple small perforations found. The tiny cytoplasmic strands which traverse the pores in such cross walls are called plasmodesmata. In the higher plants the living cells are always interconnected by plasmodesmata (Figure 1.9), so that all the living cells are actually enclosed within one plasma membrane which extends throughout the plant, enclosing the compartment known as the symplast. Because of the small dimensions of the plasmodesmata large organelles such as mitochondria, plastids or nuclei cannot pass through them, although they are frequently bridged by membranes of the endoplasmic reticulum. Large macromolecular aggregates (for example, viruses) have been shown to pass from cell to cell via this route. For a fuller discussion of plasmodesmata the reader is referred to Gunning & Robards (1976). The origin of plasmodesmata is essentially similar in fungi and higher plants, by the fusion of vesicles across the appropriate position in the hyphal segment or cell. In the case of higher plants, however, the new

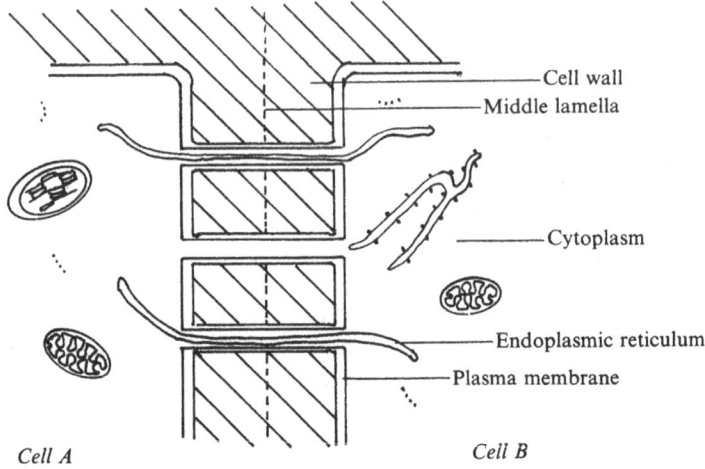

Figure 1.9. Diagrammatic representation of the plasmodesmata connecting higher plant cells together.

cross wall forms in the position of the metaphase plate of the preceding karyokinesis and grows from the centre outwards to join the peripheral cell walls of the parent cell.

The cell wall around a plant cell which is still growing is called the primary cell wall. It has layers of cellulose microfibrils, which may be randomly arranged or show some particular orientation with respect to the cell axis. There is also a matrix of hemicelluloses. This is a general name for a group of alkali soluble homo- and hetero-polysaccharides containing neutral sugars such as mannose, arabinose, xylose and glucose. There may also be a small amount of a specialized protein called extensin, which is thought to be important in providing either rigidity or extensibility to the wall, depending on the state of oxidation or reduction of its sulphydryl bonds. The outermost regions of the cell wall and the interface zones between adjacent cells contain a high proportion of a complex acidic polysaccharide material known collectively as the pectic substances. The major functional component is the sugar acid galacturonic acid. The carboxyl group may be esterified with methyl alcohol, or ionically bound to cations such as sodium, potassium or calcium. Di- and multivalent cations can form ionic bridges between adjacent molecules of the polysaccharides. The presence of pectic substances leads to a general negative charge on the cell surface and this is of significance in the immobilization of plant cells. The use of hydrolytic pectic enzymes can result in the separation (maceration) of plant tissues of an appropriate type. A slightly different mixture of enzymes, in a medium of an osmotic strength greater than that of the vacuolar sap (the liquid in the vacuole) will entirely remove the cell wall and produce protoplasts. Both of these enzymic treatments cause breakage of the plasmodesmata, which then reseal.

An obvious feature of higher plants is the extreme range of cell sizes and types which are present in a whole plant. It is the ability of individual plant cells to follow a particular pathway of structural and functional differentiation, which may be quite different from that of adjacent cells, that leads to the complexity of cell types and tissues. Certain tissues are made up of predominantly one cell type and have been used, in the form of thin slices, as part of immobilized cell biosensors (see, for example, Smit & Rechnitz, 1984). On the other hand many tissues are too complex to form a 'self-immobilized' preparation of definable activity. Conversely, plant cells grown in axenic culture are likely to have an extremely limited metabolic capability compared with those of the whole plant from which the culture was originally derived. When such cells are immobilized they are unlikely to be able to complete a normal process of differentiation and it is not yet possible to control differentiation *in vitro* to obtain a functionally homogeneous preparation of a desired cell type. Whether or not metabolic differentiation can be achieved without structural differentiation remains a topic of considerable debate.

Plant cells grown in liquid culture show a strong tendency to clump together and researchers often go to considerable lengths to provide conditions and select cell lines which do not show clumped growth. Clumps of cells may arise in two ways. Firstly, from the failure of dividing cells to separate after cytokinesis. This, of course, is the natural situation in the original plant. Secondly, cells may clump together after separation. This may arise from surface interactions, such as the creation of calcium bridges between the carboxyl groups of the pectic substances which form the outer edges of the cell wall. It is this latter type of interaction, perhaps together with hydrogen bonding and other weak intermolecular forces, which cause the plant cells to attach readily to the surface of culture vessels, particularly at the liquid–air interfaces and also to support materials intentionally added to cell suspension cultures. Even apparently 'inert' materials, such as nylon mesh (Lindsey & Yeoman, 1983) appear to attract plant cells, which become firmly attached and continue to grow and divide in the support material. The obvious consequence of this is that daughter cells will fail to separate from one another and conditions suitable for cellular differentiation are at once created. Examining the extent to which this has occurred is, however, a laborious and inherently destructive task which has not been investigated in any significant detail.

Mammalian cells

An enormous literature exists dealing with the cell biology of animal cells. For the present context it is only necessary to review briefly those features of mammalian cells which are relevant to immobilization. In the whole animal the majority of cells are self-immobilized in various tissues but some cell types are in constant motion due either to active self-movement or to passive transport in fluids. No plastids, cell walls or distinct vacuolar compartments are present in animal cells. Due to the absence of the cell wall animal cells must be maintained in an environment where the external osmotic strength is equal to, or slightly greater than, that of the internal cell fluids. In more dilute media animal cells, just like protoplasts prepared from micro-organisms or higher plant cells, will swell up and burst. This rupturing of the plasma membrane obviously destroys its ability to act as a selectively permeable membrane. It should, perhaps, be noted that under severe osmotic shock even plant or microbial cells may burst.

There are very many different types of specialized animal cells in a typical higher animal and this is clearly not the place to discuss them in detail. A generalized, if theoretical, animal cell would contain a distinct nucleus and a wide range of organelles including mitochondria, ribosomes, rough and smooth endoplasmic reticulum, Golgi systems, a pair of centrioles which are involved in nuclear division and a large number of small vesicles of various types (Figure 1.1). Among the most obvious of these vesicles are the family of lysosomes, which are rarely described from cells having cell walls. The lysosomes contain various hydrolytic enzymes which are responsible for the breakdown of material engulfed into the lysosome. The degraded material may have arisen outside the cell or from inside and, in the ultimate situation, may include other organelles essential to the functioning of the cell (Figure 1.10). When these are degraded but not replaced cell death will occur. It follows from this that although microbial and plant cells may be immobilized under harsh conditions and still retain a measure of industrially useful metabolic activity such a phenomenon is not to be expected with animal cells. Damage of any sort, even relatively mild permeabilization, may be expected to result in autodestruction of an animal cell. There is no reason to doubt that animal cells can be successfully immobilized by mild methods. The literature, however, shows an almost complete lack of interest in such mild entrapment methods as alginate. This is because attention has focused upon an entirely different aspect of immobilization, which is only to be understood by a consideration of the plasma membrane of animal cells and their behaviour in axenic culture.

The plasma membrane of animal cells is an extremely complicated

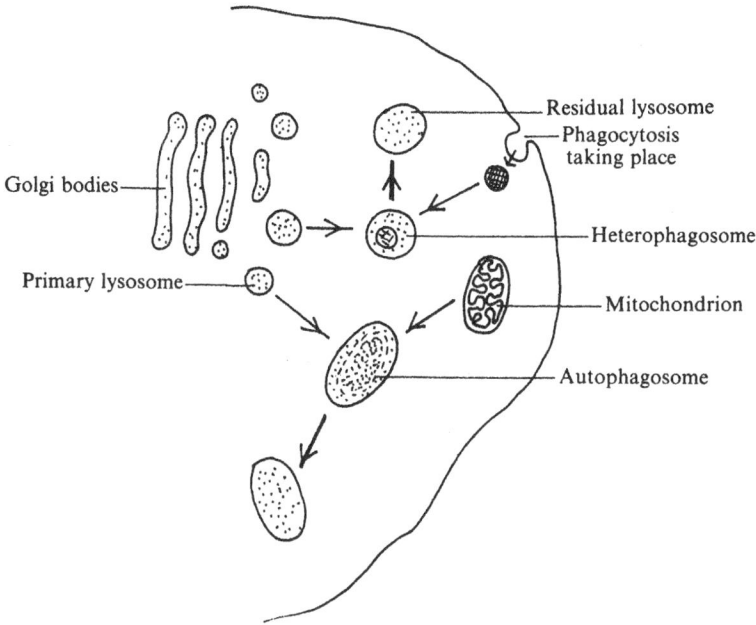

Figure 1.10. Diagrammatic representation of the lysosome cycle in an animal cell.

system controlling all types of communication with the external environment and other cells. Apart from the normal functions of a plasma membrane in controlling the entry and exit of molecules, there are also many specific communication sites. These may be receptor sites for external mobile or immobile molecules or recognition sites by which the presence of other cells, or even organisms, can be detected by the cell which carries the fixed recognition site. Such stereospecific sites are crucial to the normal functioning and behaviour of animal cells. Many, but not all, of the recognition systems involve a glycoprotein somewhere in the mechanism. Only a few types of animal cells are capable of continued growth and cell division in free suspension culture. The majority show a definitive need for a growth surface but, once that surface has been covered by a monolayer of cells, then show the phenomenon of contact inhibition and stop growing. Continued overgrowth is generally a sign of a cancerous modification of the cell. The majority of work on animal cell immobilization has, therefore, been directed towards the growth of large numbers of cells in culture vessels, either by membrane retention methods or the use of 'microcarrier' cell growth supports. These culture methods are discussed in more detail in a later chapter. The technical problems of animal cell culture mean that researchers rarely have enough cell material to carry out the type of speculative immobilization which has characterized the work on microbial and, to a lesser extent, higher plant cells. Attention with animal cells has therefore focused, for the present, upon scaling up batch culture rather than the investigation of immobilized animal cell reactors. One might reasonably expect the benefits of an immobilized animal cell reactor over the wasteful process of batch culture might justify a greater research effort but there are still many technical problems to be overcome.

The first stage of natural animal cell immobilization is obviously adhesion to the support surface. Several steps can be distinguished. Firstly, the surface must already possess or adsorb what are described as 'attachment factors'. Secondly, the cells must contact this surface and then become attached to the compatible surface. Animal cells, not being restricted by a cell wall, then flatten out and spread over the support surface, becoming more firmly attached in the process. The culture medium used for animal cell culture generally includes serum and this contains a glycoprotein called cold-insoluble globulin (CIG) which probably acts as an attachment factor. The support surface must, in addition, be hydrophilic and have a surface charge, but the charge can be either negative (as on glass and plastic) or positive (as on polylysine coated surfaces or commercially available microcarrier supports). Animal cells themselves have an overall negative charge but this is not evenly distributed. Divalent cations probably form an essential part of the attachment reaction as do proteins, which may include those secreted by the cells themselves. The name fibronectin is used for one type of cell-excreted attachment glycoprotein. Some cells secrete sufficient of these proteins to promote self-attachment to surfaces. For other cells a serum-containing medium is usually added to the culture surface before the cells. Sufficient protein is adsorbed to the surface within a few minutes. Although cell types vary in the exact conditions required for attachment the methods are essentially non-selective and rely upon a generally similar attachment mechanism (Figure 1.11).

New methods of animal cell attachment are being developed based upon biospecific (affinity) procedures (see, for example, the trade literature of Pharmacia, 1980). At the present time this is mainly being used to separate mixed cell populations, by affinity cell chromatography, but applications for the immobilization of cells producing high-value products are clearly possible. Sepharose macrobeads (250 – 350 μm diameter) have been developed as the support material. These can be surface-treated to provide various forms of biospecific reactivity. Antibodies may be attached to the macrobeads and hence cells possessing the appropriate surface antigen will be stripped out of suspension. Alternatively the antibody may be reacted separately with the cells in suspension and the cell–antibody complex specifically adsorbed onto the

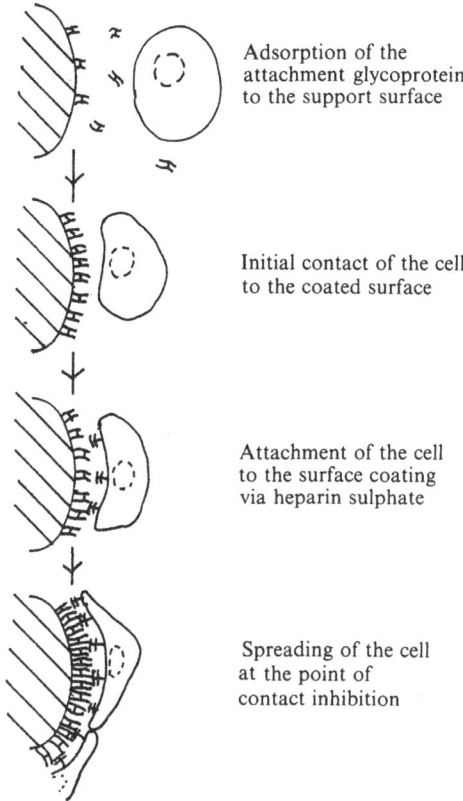

Adsorption of the
attachment glycoprotein
to the support surface

Initial contact of the cell
to the coated surface

Attachment of the cell
to the surface coating
via heparin sulphate

Spreading of the cell
at the point of
contact inhibition

Figure 1.11. Diagrammatic representation of the main stages in the natural attachment of animal cells to surfaces.

support. A second group of selection methods is based upon the reaction of Sepharose-bound lectins with cell surface glycoproteins. Thirdly, any type of cell-specific ligand can be attached to the Sepharose and used to select cells with appropriate surface receptors. Such highly selective methods may find a limited application in productive systems based upon immobilized cells but most attention will undoubtedly be given to general, non-selective, methods which can be used to immobilize a wide range of cell types.

Cell viability and cell death

Earlier in this chapter an extreme definition of cell viability was presented – the ability of a cell to divide into two daughter cells. For many practical purposes in work with immobilized cells and, indeed, often when considering cells in whole multicellular organisms, this definition is of little significance. Many cells, even though capable of division, do not actually show division. Many specialized cells are demonstrably 'living' but are capable of division only under the most exceptional circumstances, if at all. In reality, there is a continuum of cell capabilities from those that are fully functional to those in which, perhaps, only a single enzyme activity can be demonstrated. The subject matter of immobilized cell research is concerned with the totality of this continuum but the nature of the continuum is best understood by means of two alternative approaches. The first involves a discussion of the reasons for the loss of cell viability and the second, a consideration of those methods of estimating cellular function which have a direct relevance to the extent to which a cell can be considered as living.

In the introduction to the book *Cell death in biology and pathology* Bowen & Lockshin (1981) drew attention to the need for an adequate definition of cell death. They suggested that such a definition might best be sought in terms of the visible cytology of the cell rather than in the metabolic or operational functioning of the cell. This was not because they felt that cytological change came before an alteration in metabolic capability but rather that the current state of knowledge of the metabolism involved was too incomplete for precise definition. In an interesting approach to the problem they carried out a computer-aided retrieval of papers between 1971 and 1978 using the keywords 'cell death, necrosis, deletion, autolysis and apoptosis'. As might be expected from the use of such keywords, they found a majority of papers in the category of tissues from birds and mammals, because relevant work in other fields is frequently keyworded differently. Nevertheless, their survey focused clear attention on the fact that most researchers have been more concerned with the extracellular signals promoting cell death rather than the mechanism by which the cell actually died.

As Wyllie (1981) pointed out, cell death seems to involve many different structural changes but, in the case of animal cells at least, it is possible to separate those cases in which swelling of organelles or plasma membranes occurs, eventually leading to membrane rupture, from those in which a progressive shrinkage and apparent dehydration of organelles or cells occurs. The former was described as necrosis and the latter apoptosis, although Wyllie does point out that many other authors have used the terms differently. Necrosis, by Wyllie's definition, has been found primarily under conditions of cell death induced by adverse external conditions. Apoptosis, on the other hand, occurs during normal tissue turnover and embryogenesis in animals and appears to be related to a more controlled type of cell death. Energy, in the form of ATP, appears to be required and plasma membrane integrity is maintained until well into the process of apoptosis. By contrast, damage to the plasma and other membranes of the cell is characteristic of necrosis. Selective

Table 1.1. *Important aspects of the death of a
cell*

Loss of integrity of plasma membrane
Loss of integrity of internal membrane systems
Failure of specific membrane transport systems
Failure of ATP generation systems
Failure of activated coenzyme production
Failure of metabolic control systems
Failure of protein synthesis systems
Failure of lipid/phospholipid synthesis system
Failure of DNA repair mechanism
Failure of DNA replication

permeability is lost early in the process, together with the capacity to produce normal amounts of ATP. Leakage of cations and other substances in and out of the cells occurs. Cell and organelle swelling is probably associated with the failure to maintain normal osmotic concentrations and the eventual disruption of membranes finally causes loss of macromolecules and both nuclear and cytoplasmic lysis. Evidence suggests that mitochondrial ATP production is uncoupled during cell death, giving the superficially anomalous situation of increased respiratory activity but declining anabolic capability. It seems certain (see the review by Gahan, 1981) that loss of compartmentalization is a reasonable criterion of cell death and that many other detectable biochemical events, including reduced synthesis of proteins and lipids, precede this breakdown of normal membrane function. Essentially similar conclusions were reached by Roberts (1972) in a review of the cellular reasons for the loss of seed viability. Some important aspects of cell death are given in Table 1.1. From the viewpoint of immobilization it is likely that any cell which has been accidentally or intentionally permeabilized, thereby altering the normally existing membrane permeability characteristics, can be considered as technically dead. How long such a cell will remain capable of performing a useful biocatalytic function then depends more upon the stability characteristics of the enzymes being exploited than upon the specific cellular properties of the immobilized species.

Research on immobilized cells can be divided into two categories. In one the researchers have been at pains to retain 'viability', or at least the ability to regenerate active forms of coenzymes. In the other, this criterion has held no significance, because only simple biocatalytic capacities have been required and regeneration of coenzymes or *in situ* biocatalyst regeneration, by cell division, have not been required. Clearly, the latter need not concern us further at this point. Where 'viability' has needed to be assessed various criteria have been used.

Continued cell growth and division can be determined by any
conventional measure of increase in biomass, although many of these
methods are not able to distinguish between living and dead biomass.
Viable cell estimation from surface films and flocs is notoriously difficult
to achieve by conventional sampling and plate counting techniques, as
normally used for micro-organisms. The SEM has been frequently used
to show the distribution of cells on or within immobilization supports and
can, by means of time-related samples, visibly show increase in cell
numbers. With certain matrices, techniques are available for liberating
the cells, which can then be recovered and examined. In other cases the
mere process of recovery may place considerable stress upon the cells and
so lead to probable underestimates of the original *in situ* viability.

Attention has therefore focused on tests for viability which do not rely
on demonstrable cell division. Chief amongst these have been the use of
selective staining procedures. Two types can be distinguished, one relying
on the ability of an intact plasma membrane to exclude a coloured
molecule and the other on the ability of a living cell to accumulate or
modify a molecule to which even an intact plasma membrane is
permeable. In all of these tests the concentration of the 'vital' stain is
generally crucial. A substance which is non-toxic at very low
concentrations is frequently toxic if used at higher concentrations.
Among the membrane excluded dyes trypan blue and Evans Blue are
probably the most commonly used for animal cells while methylene blue
is frequently used for yeasts. Chilver, Harrison & Webb (1978) compared
23 different viable stains using yeast as the test species. It should be noted
that some confusion exists over the mechanism of differential staining
with methylene blue because some species have plasma membranes
which are permeable to it but the viable cells are capable of reducing the
methylene blue, a redox indicator, to its colourless form. Among
substances which readily penetrate the plasma membrane fluorescein
diacetate has been extensively used for viability testing of protoplasts and
plant cells. After entry into the cell the molecule is split by non-specific
esterases to yield the fluorescent molecule of fluorescein. Viable cells can
be quantified with a conventional fluorescence microscope. Leakage of
the fluorescein does occur. Many redox indicators or other substances
which change colour when reduced have been used to demonstrate the
existence of reducing capability in cells. Among these are Janus Green
and various tetrazolium salts. In general, it may be assumed that a cell
capable of reducing such exogenously supplied substrates is living.

From what has been previously said, however, it is obvious that some
enzymic activity, detectable by such reagents, may linger on well after a
cell has ceased to be living, let alone viable. Chromogenic enzyme assays
for enzymes which do not involve reduced coenzymes or ATP are clearly
even less reliable indicators of a living cell. All these staining methods

have the advantage of being both rapid and simple. For general use methods based on more detailed measurement of metabolic capability or ultrastructural changes, visible in the prepared sections using the transmission electron microscope, are far too time-consuming to be of practical significance, although they might well have research value. Functional tests of the direct capability of an immobilized cell biocatalytic system are therefore frequently the only ones which are considered worthwhile. The technicalities of determining cell viability may then be considered of only academic interest. The problem of the standardization of terminology to describe the metabolic behaviour of immobilized cells is a difficult one. Lilly (1986) presented some suggestions on behalf of the European Working Party on Applied Biocatalysts and called for comments.

2   Production and characterization of cells for use in immobilized cell reactors

Selection of cells

There have been various approaches to the selection of cells for immobilization. One has been to take an already existing species, strain or cell line, with known biocatalytic ability to synthesize or biotransform to give potentially useful products. Many different immobilization methods have then been used to determine empirically the best system for that particular cell type. A second approach has been to choose a convenient experimental cell system, such as yeast, which is readily and cheaply available in bulk and whose metabolic capability can be easily characterized. A variety of methods of immobilization have then been applied and a comparative assessment made, in general, of their benefits and limitations. A third, rather more random approach, has been to use cultures which happen to be readily available in the local laboratory and immobilization methods which, for one reason or another, also have local interest. The result of these different approaches has been a patchwork of published results which are often difficult to compare. Fortunately, in some cases, researchers have been at pains to document fully their cell material but in many others not enough taxonomic information has been provided. That this situation still continues, as indicated by the circular letter from Hawksworth to all relevant UK-based scientific journals (Hawksworth, 1984), remains a serious cause for concern. There are innumerable papers published about immobilized cells where the work cannot be adequately repeated by others due to lack of such experimental detail.

The production of High-Fructose syrups using immobilized glucose isomerase provides an example of the classical approach to the selection of species and strains for industrial uses and has been well documented by Antrim, Colilla & Schnyder (1979). Very many species, spanning over 20 different genera, were found to produce the target enzyme. From the many possible organisms only a few high-producing strains are actually used for the enzyme production. Conventional mutation-selection procedures have been applied. Several different commercially-viable production systems have resulted from this research, carried out by various industrial companies. Conventional fermenter growth has been used to obtain the biocatalyst, which has then been used in the form of

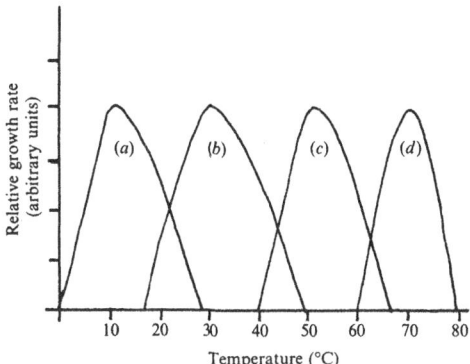

Figure 2.1. Graphical representation of the generalized relationship between growth rate and temperature for various categories of micro-organism. (*a*) Psychrophiles; (*b*) mesophiles; (*c*) thermophiles; (*d*) extreme thermophiles.

either immobilized enzymes or immobilized cells. In the case of glucose isomerase it is perfectly satisfactory to use cells which are dead because only a simple, single-step, non-coenzyme-requiring reaction is involved. Thus cells adsorbed onto porous supports and anion-exchange resins, glutaraldehyde cross-linked (with and without added polymers) and even merely heat-fixed cells, have all been used. This also means that all required biocatalytic activity must be generated prior to immobilization. Because many enzymes require positive induction the composition of the culture medium is often crucial. In the case of glucose isomerase, which is actually xylose isomerase in terms of its major activity, the inducer xylose is too expensive for bulk use. Cheaper, crude inducers were used at first and then constitutive strains which require no inducer. Because the successful operation of an immobilized cell reactor depends on the stability of the biocatalyst, an obvious development is always to seek more thermostable organisms. In the case of glucose isomerase a further advantage arises because the equilibrium concentration of fructose increases with rise in temperature. In general, it has not proved easy to create thermophilic mutants of mesophilic organisms, but there have been numerous successful searches for constitutive mutants capable of synthesizing target enzymes without the need for induction. The general experimental approach in such cases is therefore, whenever possible, to use thermophilic micro-organisms (Figure 2.1) originally isolated from appropriate natural environments. It is also often found that the enzymes from the thermophilic micro-organisms are also more stable to organic solvents and other denaturants and can therefore reasonably be expected to show a higher expressed activity after immobilization. De Rosa *et al.* (1980) used this approach when seeking a successful system for lactose hydrolysis. They found that cells of the extreme thermophile, *Calderiella acidophila* could be cultured at 87 °C at pH 3.0 and showed constitutive β-galactosidase activity. Cells immobilized in polyacrylamide still showed extremely high retention of β-galactosidase activity (assayed at 80 °C). The activity after immobilization was much greater than that expressed by intact cells and comparable with that of acetone-permeabilized cells. Stored in a wet form, at 4 °C, the polyacrylamide-entrapped cells retained full β-galactosidase activity for at least eight months. The half-life of activity at 70 °C was ten days. The cells were not viable after polyacrylamide entrapment but the β-galactosidase activity was stable to treatment with organic solvents (necessary to remove milk fats under proposed commercial applications). The very high temperature of operation would be likely to prevent unwanted microbial contamination.

Starting with the thermoacidophilic archaebacterium *Sulfolobus solfataricus*, Nicolaus *et al.* (1986) were able to use a very harsh immobilization treatment and still retain 2-keto-3-deoxygluconate production from gluconate. Competing aldolase activity in the preparation was inhibited by sodium borohydride treatment. This is an example of the use of thermophilic organisms to obtain stable biocatalytic activity together with the non-genetical manipulation of the spectrum of enzymes. The latter approach may be the only way to prevent competing reactions which are part of the central metabolic pathways essential to the metabolism of all organisms.

It is not always necessary for an organism to be itself thermophilic for it to contain a specified thermophilic target enzyme. It is a common observation that virtually all organisms have enzyme complements which show differing sensitivities to heat denaturation so that the optimum growth temperature may be well below that of a single chosen target enzyme. Unfortunately, the reverse situation can also apply for the more complicated biosynthetic capacities. In their classical paper on the growth, respiration and penicillin production of *Penicillium chrysogenum*, Calam, Driver & Bowers (1951) showed that although the optimal temperature for growth was around 30 °C, the production of penicillin was optimal at just under 25 °C. The existence of such relationships is yet another reason for attempting to separate growth of an organism from production of a desired product, an end to which immobilization provides an excellent route.

Conventional fermentation processes

Because an immobilized cell reactor represents, in many cases, an alternative to conventional fermentations, it is convenient at this point to survey selected aspects of such fermentations. This also provides a basis for the operational design of immobilized cell reactors which is developed in Chapter 7. It should also be remembered that, with very few exceptions, cells destined for immobilization have first to be grown in and harvested from conventional fermenters. The only real exceptions to this are small-scale devices using tissue slices from multicellular organisms and the categories of fermenters which have been specifically designed to maximize attached microbial or other cellular biomass. In the latter type there is an initial fermentation stage, during which active biomass is built up, which can then be followed by a period of biocatalytic activity with minimal (or no) cell growth.

The kinetic patterns of fermentations have been classified in various ways which emphasize different features of conventional fermentations. Gaden (1955) proposed three types of fermentation based essentially on the nature of the relationship between product formation and the utilization of the carbon source for metabolism (in his system specified as carbohydrate).

Type I, of which ethanol is an example, shows a direct relationship.

Type II, with citric acid as the example, shows an indirect relationship.

Type III, with penicillin as the example, shows no obvious relationship between product formation and carbon source utilization.

Viewed biochemically, rather than in terms of simple quantitative relationships, these same examples can be seen to illustrate other features of fermentations. An alternative classification has grown up which distinguishes so-called 'primary' products, which are substances directly related to the primary metabolic pathways essential for virtually all cells, from 'secondary' or 'species-specific' products which are synthesized only by a limited number of species. The secondary products are generally produced only in the later stages of a fermentation or in differentiated cell types of multicellular organisms. In Gaden's (1955) examples ethanol and citric acid are primary products and penicillin is a secondary product. Even this view is obviously simplistic when the biochemistry is taken further. In ordinary heterotrophic growth the growth substrate has two roles. One is to provide the carbon atoms which are the basis of the molecules produced by the cell and the other is to act as a source of energy to allow this biosynthesis to proceed. During cell growth the majority of the cell's energy need is for the synthesis of enzymic and other proteins. Metabolic control systems, of which there are many different types, ensure that as long as conditions remain favourable growth and cell division continue. Naturally, or by purposeful manipulation of the culture medium and environment, a particular essential nutrient generally becomes limiting. Depending on the limiting nutrient (for example, carbon source, nitrogen source, phosphate or other inorganic nutrient) the metabolism may become directed towards particular products which are of potential commercial significance. An alternative cause of metabolic differentiation is the genetically pre-programmed sequence of changes which occurs during cellular differentiation.

It is also possible to classify products as growth-related (roughly equivalent to primary products) and non-growth-related, but it is important to remember that even when growth and cell division have ceased a certain proportion of the cell's energy must be devoted to 'maintenance'. This corresponds to the requirements for the synthesis of essential membrane components and other molecules. This maintenance energy is also needed when cells are immobilized in the living state but is not, of course, relevant to non-living immobilized cells. A distinction also needs to be made between products which are the result of total synthesis and those that represent only a biotransformation of an added molecule and where the product therefore retains a considerable proportion (if not all) of a molecule originally supplied to the cell. To clarify this point a greater depth of knowledge of cellular metabolism and the flow of carbon in such metabolism is needed. Returning briefly to our three products ethanol, citric acid and penicillin, it is well-known that ethanol represents a relatively intact part of the glucose molecule which is the starting-point of the process of glycolysis. Of the six carbon atoms in a glucose molecule the pathway of catabolism via two interconvertible three-carbon molecules means that only two molecules of ethanol can be produced and not three as might at first sight seem possible. The remaining two carbon atoms are liberated as carbon dioxide and only a small amount of the chemical potential energy present in the glucose is actually available to the alcohol-producing cell to support its growth. The bulk of the energy content of the glucose remains in the ethanol, which is lost from the cell compartment. This is the situation in yeast under anaerobic conditions.

In other micro-organisms different routes to alcohol production may be used. For citric acid the metabolic pathways involved are more complicated and it is rather less easy to quantify the amount of citric acid that should theoretically be produced if the metabolism were directed efficiently towards that end. In the case of penicillin this is even more complex and, in practice, the diversion of carbon atoms into the penicillin molecules is low, with yields of the antibiotic being only of the order of a few per cent (Figure 2.2). In cases like the latter the possibility of biotransformation rather than total *de novo* synthesis is realistic and major subunits may be fed to the culture for more efficient incorporation. At its extreme, as in certain steroid transformations, a complex substance may be fed to a cell with the aim of obtaining only very minor, but economically significant, changes to its structure. Because many enzymes are only partially stereospecific it is even possible to feed substances which are not naturally synthesized by the species being used for biotransformation.

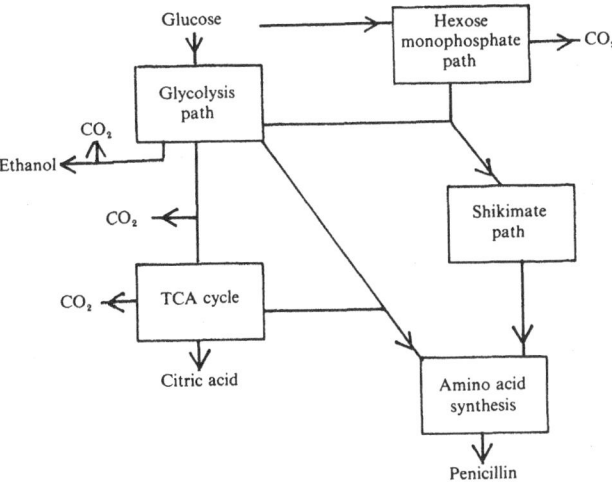

Figure 2.2. Simplified diagram of the interrelationship of the central pathways of metabolism leading to product formation.

One of the obvious advantages of immobilized cells over immobilized purified enzymes is their potential ability to carry out complex biocatalytic reaction sequences in the same way as living free cells in a fermenter. Vieth & Venkatsubramanian (1979) gave a brief discussion of this topic but described it as being 'still in its infancy' at that time. Welch (1978) considered the alternative of creating covalently linked multicomponent immobilized enzyme systems as unlikely to develop significantly due to the problems of optimizing such systems. The use of immobilized cells to synthesize the total molecular structure of coenzymes, not just the active form of coenzymes as required to promote complex biocatalytic reactions, has received some attention. Shimizu, Tani & Yamada (1979), for example, compared the synthesis of coenzyme A by *Brevibacterium ammoniagenes* by fermentation and immobilized cell methods and concluded that the latter was a simple and practical process. As might be expected considerable attention has been given to the possibility of producing antibiotics by immobilized cell processes. As early as 1979 Suzuki & Karube (1979b) described the synthesis of penicillin G and bacitracin by immobilized cells but at the present time (1986) it is still true to say that the majority of antibiotics are produced by conventional fermentations. Vandamme (1983) reviewed the production of peptide antibiotics by immobilized cells, pointing out the great potential of such systems for replacing conventional fermentations. The complexity and capital investment profile of the antibiotic production plants already in existence undoubtedly militates against drastic changes in production processes.

Biocatalytic activity

The biocatalytic activity of cells and cell preparations resides in the enzymes present in them. The precise order of amino acids in the polypeptide chains making up an enzyme is predetermined by the information contained in the cellular DNA and constitutes the primary structure of the enzyme. Depending on the type and order of amino acids a secondary structure is also predetermined, in which parts of the polypeptide chains may take up helical, random or pleated sheet arrangements. Further folding due to ionic, covalent, and hydrogen bonding and hydrophobic–hydrophilic interactions gives rise to a tertiary structure. The structure of the active enzyme may be further complicated by the aggregation of complete polypeptide units into a quaternary structure. In simple enzymes the predetermined three-dimensional structure of the protein determines the enzyme's specificity. Only a relatively small part of the protein actually forms the active site, which is the site at which substrate is converted to product. Changes in the other parts of the protein may have little effect on biocatalytic activity. In nature, families of enzymes (isoenzymes) exist that have essentially the same catalytic action but which differ in their more subtle properties. A single species may be capable of producing more than one isoenzyme, either in different tissues and organs (for the higher eukaryotes) or within a single cell. Because these isoenzymes may be inhibited to varying extents by metabolites arising in the cells, it is possible for them to act as control stages in branched metabolic pathways.

Perhaps the most widespread example of metabolic control at the enzyme level is the presence of allosteric enzymes (Figure 2.3) at the start of metabolic pathways. In addition to the active site such enzymes have binding sites for allosteric inhibitors and/or activators (effectors). Products of enzymes which occur further along the metabolic pathway can bind at these sites and slow down, or shut off entirely, the particular reaction sequence. This is a rapid way of shutting down a whole pathway without the multiple steps required if end-product inhibition of each individual enzyme had to take place sequentially.

A clear distinction must be drawn between this type of inhibition and that caused by a lack of the active forms of essential coenzymes and/or cofactors, which affect the reaction at the active site itself. Much of the metabolism of cells is, of course, influenced by the supply of an energy source which often (but not inevitably) also acts as a carbon source for biosynthesis. These two roles, although linked together, are alternatives. If glucose, for example, is being used to generate the active forms of coenzymes such as ATP and $NADH \cdot H^+$, under aerobic conditions, the carbon atoms will be completely oxidized to the carbon dioxide that is liberated during cellular respiration. As much as 60% of the carbon will

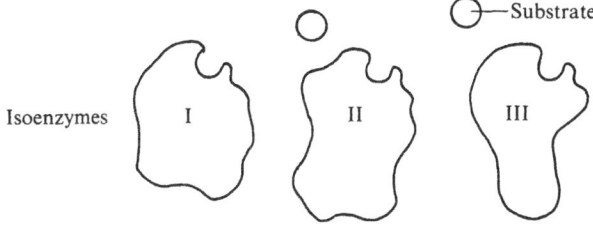

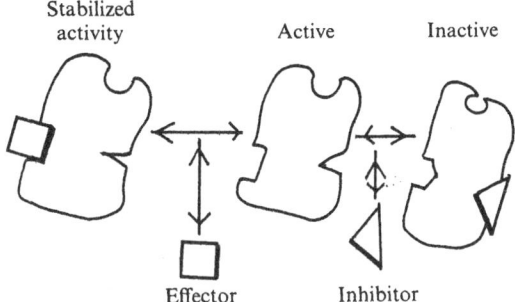

Figure 2.3. Diagrammatic representation of three isoenzymes (I, II, III) which have the same biocatalytic activity but differ in amino acid sequence, and an allosteric enzyme which can be active or inactive according to the presence of effectors or inhibitors.

need to be liberated in this form to provide sufficient active coenzymes to metabolize the remaining carbon atoms into new biomass or complex products. Biomass increase, under defined conditions, can be predicted simply from carbon dioxide evolution. Shifting from aerobic to anaerobic conditions causes a greater flow of carbon to carbon dioxide because anaerobic respiration is a less efficient way of generating active forms of coenzymes. Incidentally, useful products such as alcohol may be liberated. The alcohol still contains a considerable amount of the energy originally present in the molecular structure of the glucose. Ellwood *et al.* (1982) proposed a theory based on the widely accepted chemiosmotic principle of ATP generation in membrane systems to account for the greater growth of immobolized cells compared to those in a free culture. This was based on increases in the proton concentration at sites of immobilization (Figure 2.4).

To characterize the biocatalytic activity of immobilized cells and to compare this with the biocatalytic activity of the same cells in a free state clearly requires convenient quantitative measures of such activity. In conventional enzymology the procedure is to develop a quantitative assay

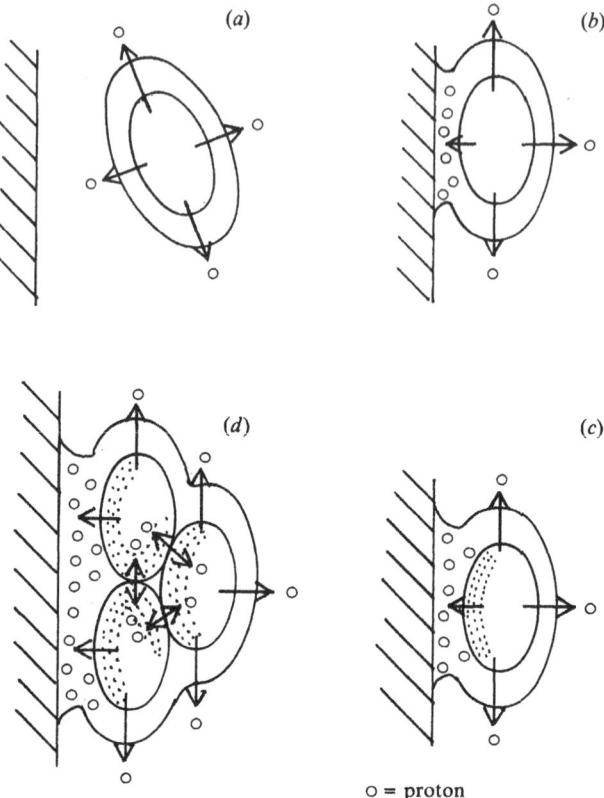

Figure 2.4. Diagrammatic representation of the chemiosmotic theory of metabolic activation resulting from immobilization. (*a*) Free cell extruding protons; (*b*) protons accumulating in the cell–surface interface; (*c*) localized polarization of cell activity on the interface side; (*d*) microcolony sharing proton gradients with more widespread promotion of metabolic activity. (After Ellwood *et al.*, 1982.)

for the enzyme being studied and to attempt a total extraction of the active enzyme or, at the least, a known percentage recovery, to enable the total potential enzyme in the cells to be quantified. Appropriate units of enzyme activity (which are ideally expressed in SI units but in practice are often not) per unit of extracted protein or original biomass or cell number can then be calculated. The actual expressed activity of the free cells might be much less than this theoretical maximum value due to the operation of intracellular control systems or the restriction of substrate diffusion into the cells and/or product diffusion out of the cells. This leads to the concept of 'expressed activity' rather than true or potential biocatalytic activity. Again the expressed activity may be quoted per unit of protein, biomass or cell number. In the case of immobilized cell preparations the general validity of this approach is recognized and a preparation can be reasonably described by its expressed activity, under defined assay conditions, in appropriate units of biocatalytic activity. Because the assay of protein content, biomass or cell number is often difficult, if not almost impossible, the activity of immobilized cells is more often expressed on a basis of fresh or dry weight or volume of the immobilized cell preparation. This at least defines, in an acceptable way, the starting value of an immobilized cell preparation.

In practical studies of immobilized enzyme and cell reactors the biocatalytic activity is usually expressed in percentage terms relative to the activity of the free enzyme or cell preparation. Both assays are carried out under similar conditions of substrate concentration, pH value and temperature. In some cases, e.g. the proposed guidelines for the characterization of immobilized enzymes which were included, to stimulate discussion, in Broun, Manecke & Wingard (1978), the term effectiveness is also used in the same way as efficiency. There are, however, specific uses of the term effectiveness factor (see Atkinson, 1974; Halwachs, 1979; Luong, 1983; Venkatsubramanian, Karkare & Vieth, 1983, for various discussions and explanatory approaches to this) and it is better to avoid any chance of confusion between simple experimentally determined comparisons of free and immobilized cells, which can be considered as a valid use of the term efficiency and the much more complicated definitions and usage of the term Effectiveness Factor. It is at this point that the uninitiated researcher will begin to find the going rather difficult. In their desire to find generally valid relationships, biochemical engineers have produced a wealth of equations enabling the calculation of the effectiveness factor under a variety of situations. Much of the work already done for immobilized enzymes (for example, Engasser & Horvath, 1976) is applicable to immobilized cells. Even where a multi-enzyme system is involved the kinetics of the overall process are often limited by the activity of one particular key enzyme in the reaction chain and hence relatively simple enzyme kinetics can be used. The only point to bear in mind is that if the cells are still viable growth may occur and change many of the components in a mathematical equation. Most authors agree that the Greek symbol η should be used to designate the effectiveness factor, that it is dimensionless and bears an inverse relationship to the Thiele modulus, which is itself described as a dimensionless group.

Operational life of a biocatalyst can be considered in two ways. Firstly, because the loss of biocatalytic activity is essentially a random form of molecular degradation, leading to loss of the active conformation of the

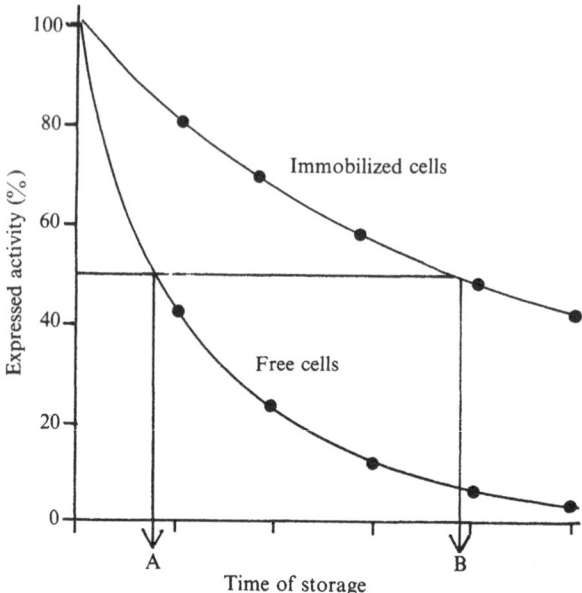

Figure 2.5. Graphical representation of the concept of the half-life of biocatalytic activity of free (A) and immobilized (B) cells. Cells stored under a specified set of conditions of temperature, pH value and substrate concentration.

enzyme, it can be treated in the same way as radioactive decay. This leads directly to the concept of the half-life (Figure 2.5) of an immobilized cell preparation as a comparative measure of different preparations and in comparison with free cells. Note that it is not a case of comparing the amount of activity based on the original potential content of the free cells, but instead on the original expressed activity of both the free cells and the immobilized cells. In this way the loss of activity which generally occurs during the immobilization process does not complicate the comparisons. The only problem remaining is that different storage conditions are used by different workers and for different cell preparations. Comparisons between preparations are therefore rarely quantifiable in an exact manner. The second approach is to define stability in terms of operational process cycles (Figure 2.6), which may show varying profiles of temperature, pH value and other conditions. This is perhaps commercially more valuable because it could give a measure of practical performance.

As part of the initial academic research into any system of immobilization it is also usual to obtain data on certain fundamental characteristics of the expressed biocatalytic activity. This is generally

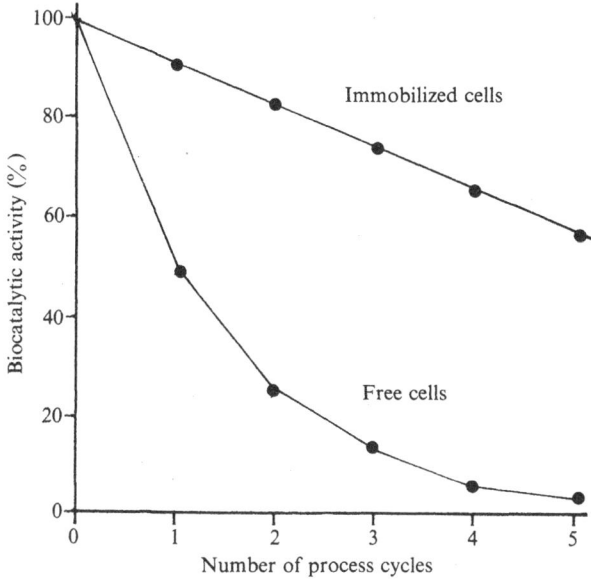

Figure 2.6. Graphical representation of the concept of the operational life of a biocatalyst in terms of number of process cycles completed. Biocatalytic activity is determined at the start of each process cycle. During an individual process cycle the operational conditions may change markedly in terms of temperature, pH value and substrate or product concentration.

treated in terms of individual target enzymes rather than multi-enzyme systems. Most frequently the influence of substrate concentration upon expressed activity is determined and from this is derived, generally by graphical means (such as the Lineweaver–Burk method) a value of apparent Michaelis constant, K_m app. The question of the nomenclature (terminology) to be used in discussions of immobilized biocatalysts was briefly considered by Sundaram & Pye (1978b). K_m is a measure of the affinity of an enzyme for the substrate under study and assumes free enzyme, operating under homogeneous conditions and without diffusion limitation. As might be expected, immobilized preparations of cells and enzymes almost invariably show an apparent reduced affinity of the enzyme for its substrate and hence an increased numerical value of K_m. This is probably mainly due to diffusion limitation but may in part relate to alterations in the three-dimensional structure of the enzyme as a result of immobilization procedures (Figure 2.7). The former effects may be partly overcome by careful attention to the size and shape of the support materials and the selection of an efficiently designed reactor configuration. A detailed discussion of the effect of temperature upon the

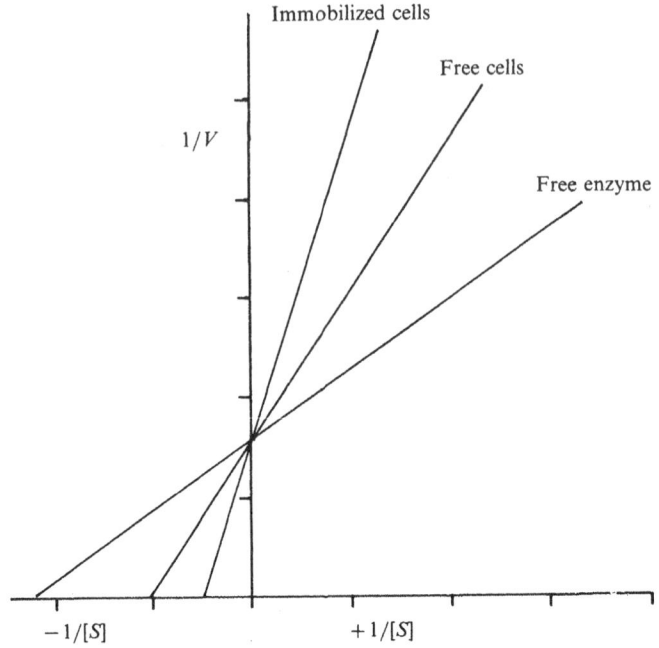

Figure 2.7. Comparison of the true Michaelis constant of a free enzyme with the apparent Michaelis constants of the same enzyme in free and immobilized cell preparations (Lineweaver–Burk plots). The actual relationship is more complex than is illustrated here and the lines for the cells are only approximations to the true situation. V = initial velocity of reaction; $[S]$ = substrate concentration.

activity of glucose isomerase in gelatin-entrapped whole cells of *Actinoplanes missouriensis* is given by Roels & Van Tilberg (1979).

The influence of both intrinsic enzyme characteristics and diffusion limitation are contained within the Thiele modulus ϕ. This is variously expressed by different authors but is probably most easily understood in the form:

$$\phi = L[K_{true}S_s^{m-1}/D_e]^{0.5}$$

where L = characteristic dimension of the immobilized cell preparation, K_{true} = the true (intrinsic) kinetic constant for the biocatalytic activity of the cells, with no masking influences, S_s = substrate concentration at the surface of the immobilized cell preparation, m = the order of the reaction being catalysed, D_e = the effective diffusivity.

Some of the mathematical analyses of the activity of immobilized call preparations have been based upon zero order reaction kinetics, which implies that the reaction rate is independent of the concentration of the substrate. This does not occur very often except where other factors are limiting and most analyses are based on first order kinetics. For almost all practical purposes this is quite sufficient. Even if several enzymes are involved in a complex reaction pathway the overall rate of reaction for the pathway will be limited by the rate of reaction of the slowest component. Likewise there is no need to draw a distinction between unimolecular and pseudounimolecular reactions. In a hydrolytic reaction, for example, the concentration of water present will normally be so high that it can be assumed to be constant throughout the reaction.

Two obvious conclusions can be drawn from the above equation:
1. The most influential component is L, because ϕ is directly related to its value but only to the square root of the remaining function.
2. Reducing the ease with which substrate reaches the immobilized cells will increase the value of ϕ.

From what has been said about the Michaelis constant it will be clear that the extent to which the immobilized cell is saturated with substrate (in effect the ratio of K over S_s, the substrate concentration at the site of reaction) will be important. Enzymes fully saturated with substrate will obviously be capable of expressing their full biocatalytic activity, other factors being non-limiting. Values of $\phi \leq 0.1$ can be considered to indicate negligible diffusional limitation. As the value of ϕ increases so does the influence of diffusional limitation.

The effectiveness factor η, which allows comparison of the efficiencies of immobilized cells with free cells is another dimensionless factor.

For spherical particles in a packed bed:

$$\eta = [1/\phi(1/\tanh\phi - 1/3\phi)]$$

(Venkatsubramanian *et al.*, 1983)

For systems showing Michaelis–Menten type kinetics a modified ϕ' is required in which there is separate consideration of terms which depend on or are independent of the concentration.

This takes the form:

$$\phi' = L[V_{app}/K_m appD_e]^{0.5}$$

where V_{app} = apparent maximum velocity, $K_m app$ = apparent Michaelis–Menten constant.

This has been explored by Halwachs (1979). The interested reader is referred to a major contribution on the kinetics and diffusion effects by Engasser & Horvath (1976). Although this deals with immobilized enzymes, much of what is presented is broadly relevant to immobilized cells. The contributions of Venkatsubramanian *et al.* (1983) and Luong (1983) should also be consulted. The relationship between η and ϕ is shown in Figure 2.8.

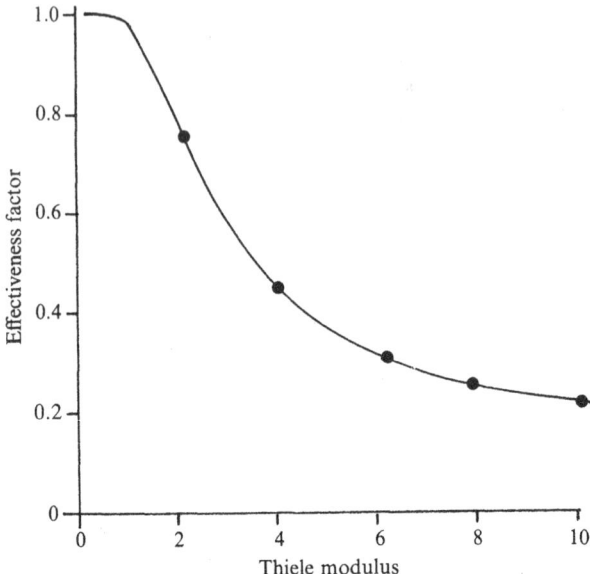

Figure 2.8. Graphical representation of the relationship between the Thiele modulus and the effectiveness factor (after Venkatsubramanian, Karkare & Vieth, 1983). The shape of the graph varies considerably for different types of particles and often does not show the steep drop illustrated here for first-order type kinetics and rectangular particles.

Growth of cells for immobilization

In general, no special conditions are necessary for the growth of cells which are to be immobilized compared with those for normal fermentations. Because the fundamental principle behind immobilization is to extend the functional life of the biocatalyst, generally under non-growth conditions, there is little point in substituting an immobilized cell reactor for a process which operates efficiently in a continuous fermenter. Most cells for immobilization are therefore grown in batch fermenters in the conventional manner, harvested at a fairly advanced stage of the fermentation, such as late linear phase, and then immobilized. This is generally sufficient to release the cells from the catabolite repression which frequently occurs during rapid cell growth and hence promotes the formation of certain commercially useful metabolic products. In other cases earlier harvesting may be required to give maximum activity in the immobilized cells, as, for example, the β-glucosidase activity of *Trichoderma* E58 in alginate beads (Matteau & Saddler, 1982b). There is obviously little point in repeating here what can be found in any good book on microbial growth in fermenters, such as Pirt

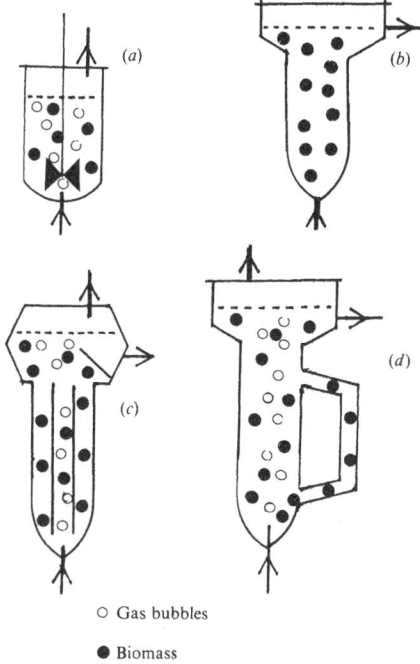

○ Gas bubbles

● Biomass

Figure 2.9. Diagrammatic representations of some commonly used types of fermenter. (*a*) Stirred-tank; (*b*) fluidized-bed; (*c*) air-lift with internal draught tube; (*d*) air-lift with external recirculation loop.

(1975). This deals primarily with conventional microbial batch and continuous stirred tank reactor (STR) fermenters, as do Aiba, Humphrey & Millis (1973). A wider range of reactors and fermenters is considered by Atkinson (1974). There has been considerable development of what are described as new or novel fermenter configurations (see, for example, Katinger, 1977). Some commonly used reactors are shown in Figure 2.9. The culture of fragile cells (essentially those from multicellular animals and plants which show extensive structural and physiological differentiation) has progressed rapidly although problems still remain which are inherent to the cells being cultured. Pollard & Khosrovi (1978) discussed some principles of reactor design for such cells and gave special attention to a tubular flow reactor used for the production of a polyhedrosis virus of *Autographa californica* in cells of *Tricoplusca ni*. For plant cells considerable debate still exists over the advantages of air-lift fermenters over conventional STR systems. The early work showing increased productivity in air-lift systems, such as that for anthraquinones from cultures of *Morinda citrifolia* (Wagner & Vogelmann, 1977) has not always been found with other cell types. It is worth noting here that in the subject area of plant cell fermentations the terms Type I and Type II refer, respectively, to cell cultures which grow mainly as single cells and those which grow mostly as multicellular flocs or pellets. This is a quite distinct use of the terms to that noted earlier for microbial fermentation types. The general conclusion seems to be that, while the air-lift systems, in whatever form adopted, have probably the widest application, it is possible to select, by repeated batch or by continuous operation, cell lines which can grow in STR fermenters, provided these are operated within certain rather critical limits of aeration and stirring. Often modified stirrer configurations are needed. In the case of animal cells little attention has been given to air-lift systems. Animal cells in culture are broadly classified into those which require a surface support for growth and those (generally transformed or cancerous cells) which can grow freely without a support surface. The former cells generally show contact inhibition with respect to other cells and grow as monolayers. The latter type can overgrow one another. For cells which grow only as monolayers a specialist area of immobilized cell technology has developed. This takes two quite different approaches to the problem. In one the surface area available for growth is increased by the use of microcarrier beads (see, for example, Pharmacia, 1981). These can be used in slightly modified STR systems. In the other, hollow fibres are used in the manner of artificial capillaries, in specially designed culture systems (see, for example, Amicon, 1985).

An interesting point which relates to the form of cells for immobilization arises when filamentous organisms are being cultured. One approach is to disrupt the filaments after harvesting and before immobilization. Deo & Gaucher (1983) using *Penicillium urticae* for the production of patulin compared intact, broken and homogenized mycelia and then allowed the K-carrageenan-immobilized fragments to regrow during a 48 hour incubation before use. In this case the final immobilized preparation from the fragmented mycelia actually showed a longer half-life than the intact mycelia. An alternative approach used by Livernoche, Jurasek, Desrochers & Veliky (1981) was to culture the fungus *Coriolus versicolor* in flasks containing a large (2 cm diameter) glass bead on a shaker at 250 rpm which kept the fungus in a finely divided form. The final immobilization matrix in this case was alginate and the preparation was used for the successful decolorization of papermill effluent.

Cells requiring special cultural conditions will necessitate special immobilization procedures. Jones, Guyot & Wolfe (1984), for example, had to take special precautions when immobilizing defined consortia of methanogenic anaerobes in agar gels. In this case methane generation was similar for both free and immobilized consortia. The small pellets (2.5 mm × 2.5 mm × 6 mm) used in this work do not appear to have caused problems of gaseous product diffusion. They do illustrate, however, the distinct advantages which can arise when dealing with mixtures of microbial species which act together. In a conventional batch or continuous fermenter it may be very difficult to maintain an appropriate balance of the populations of each of the species concerned. With immobilized consortia the desired mixture can be prepared initially before immobilization and the reactor then operated in a continuous mode without serious danger of the wash-out of individual species. In the case of methanogenesis close proximity of the species in the consortia is required. In other cases separate immobilization of each species may be an adequate approach, with subsequent mixture in one vessel to give a completely mixed reactor or sequential use in a plug-flow process.

The question of activation of metabolic activity prior to immobilization is complex and the apparent induction of increased enzyme activity may not always prove to be genuine. Chibata & Tosa (1976), for example, found that the 10-fold increase in aspartase activity of immobilized *E. coli* cells was due to autolysis and removal of diffusion limitation rather than new protein synthesis. Genuine examples of pre-immobilization induction are also frequent in the literature. Frein, Montenecourt & Eveleigh (1982), for example, used lactose to pre-induce the cellulase of *Trichoderma reesei* before immobilization in K-carrageenan. The immobilized cells continued to produce cellulase for many days, although maximum productivity of cellulase active against filter paper occurred on day four of continuous operation of a fluidized bed reactor. In an unusual application Khachatourians *et al.* (1982) induced thymine phosphorylase activity by the addition of thymidine before agarose gel immobilization of 'anucleate' minicells of *E. coli*, which are produced in certain strains by unequal cell division. Induction before immobilization is clearly essential here.

Techniques for studying immobilized cells

A detailed consideration of this topic would fill a complete book (see, for example, Buchholz, 1979) and only a brief discussion is possible here. The target biocatalytic activity of an immobilized cell preparation can be determined in two ways:

Firstly, by the quantitative assay of the target enzyme(s) using standard assay conditions in the same way as for free enzymes. The conditions of the assay may bear little relationship to the operational environment of the immobilized cell preparation but they are a convenient way of describing activity for marketing or comparative purposes. Often a convenient alternative substrate is used in place of the true target substrate. The former might be chromogenic, or more easily handled due to better physico-chemical properties. It is desirable to express enzyme activity in terms of bonds broken (for example μmoles of a particular covalent bond type) but this is frequently not convenient or even impossible. The results of a standard enzyme assay may therefore be reported in terms of substrate mass decrease, product mass formed or even arbitrary units which cannot easily be related to any of these measurements. For details of particular assays the reader is referred to appropriate textbooks and papers.

Secondly, by the quantitative assay of the target biocatalytic activity under some specified operational conditions which are appropriate for the application of the immobilized cell preparation. Researchers rarely choose the same conditions and comparisons between preparations are frequently uncertain, if not impossible, unless carried out within one particular published piece of research. The reader should consult appropriate papers for details of assays for the biocatalytic activity of specific immobilized cells.

The visual examination of immobilized cell preparations is considered to be an important source of information on the physical relationship of the cells to the support or gel matrix material used for immobilization. The use of the light microscope for this purpose is infrequent due to the limitations of magnification and particularly its generally inadequate depth of forms. The most commonly used instrument is the scanning electron microscope (SEM) which offers the possibility of variable magnification up to an amount which is more than sufficient even for bacterial cells. The great depth of focus allows porous gels and support materials to be easily examined (and photographed). Being essentially a surface examination technique it is applicable to all types of support material, even those which are completely opaque to both light and electrons. Although observation of fresh immobilized cells, in the frozen state on cryogenic stages, is possible, this simple technique is limited by lack of availability of the facility to many researchers. Generally, semi-permanent preparations are prepared by critical point drying of the specimen on a standard SEM specimen stub and evaporative coating with an appropriate electron-conducting layer. The use of the SEM is a specialist technology and is generally carried out by specially trained personnel rather than the researchers actually involved in the immobilization procedures. Specialist books should be consulted for further information on this topic.

The transmission electron microscope (TEM) is less frequently used than the SEM. Its major role is in the examination of sections of cells immobilized in or on organic supports which can be sectioned in the same way as the cells themselves. Careful fixation, staining, embedding and sectioning are necessary, again carried out by specialist staff. The major use of this more time-consuming specimen preparation has been to demonstrate the intact or degraded state of cell membranes, after

Table 2.1. *Some commonly used primary methods of examining immobilized cells*

Technique	Advantages	Limitation
SEM	Relatively easy and visually shows relationships of cells to each other and support materials	Surface technique only Gives no indication of metabolic state
TEM	Shows structure of membrane systems and organelles	Involves complex techniques Difficult or impossible for many inorganic supports
Fresh weight	Simple, rapid method	Water content may be influenced by conditions Not easy with water-holding supports
Dry weight	Simple method	Dry weight may not relate in a simple way to cell number
Cell counting	Simple, rapid method	Cannot be done directly unless cells can be released from support material

immobilization or deliberate permeabilization. By appropriate techniques it is possible to investigate the localization of particular metabolic reactions but this has been rarely attempted with immobilized cell preparations. The reader is again referred to standard textbooks for such methods. Some methods for examining cells are given in Table 2.1.

The determination of the fresh or dry mass (weight) of cells in an immobilized preparation is a serious problem and each particular type of preparation and situation must be considered separately. The method of Uribelarrea, Pacaud & Goma (1985) may well prove useful in the practical determination of the water content of immobilized cells. They used an analysis of drying curves of samples on an infra-red balance to allow the rapid determination of internal and external water contents. No doubt the presence of extracellular gels would complicate the analysis, but with microcomputer linked data processing this should not be an insurmountable problem. Determination at the time of immobilization is in general relatively straightforward. The quantity of cells before immobilization can be readily determined by sampling. In the case of cells embedded in gel matrices or behind membranes, virtually all the cells can be entrapped. With surface support materials any cells remaining in the washings after immobilization can be recovered and the mass measured. In cases where cell number rather than mass is required, samples can be enumerated before immobilization and in any washings after immobilization. This can be done by simple light microscopy using standard counting grid slides such as the various haemocytometers which

Table 2.2. *Some commonly used secondary methods of examining immobilized cells*

Technique	Advantages	Limitation
Protein content	Reliable, simple, sensitive	Cannot be used if support material or immobilization method interferes in assay
Nucleic acid content	Reliable, sensitive	Assay subject to interference and relatively complicated
Carbohydrate content	Reliable, sensitive	Difficult to relate to active biomass, subject to interference by some supports and metabolically variable

are available. Where an automated cell counting system, such as a Coulter Counter, is available this can be done both quickly and conveniently. The simplest techniques therefore involve a calculation of differences before and after immobilization. Direct measurement of cells in an entrapped preparation is possible if the gel matrix, as in the case of alginate, can be reconverted to a completely soluble form and so liberate the cells again. With non-covalent surface adsorption it is also often possible to release the cells by adjustment of the pH value or ionic strength. This may only be possible immediately after adsorption because cell growth or metabolism may result in much stronger adhesion to the surface support. The amount of leakage of cells from a support is an important characteristic of any immobilization system, and methods capable of detecting small numbers of cells, such as direct cell counting, are then useful. Cell leakage may arise from loss of the same cells as were originally immobilized or from subsequent cell division and loss of the unattached daughter cells.

Because the direct determination of cell mass or number may be difficult or time-consuming, many researchers prefer to cross-calibrate some particular organic component of the cells against fresh or dry weight and then use a biochemical method of analysis (Table 2.2). With support materials of sufficiently different composition to the cells it is possible to determine accurately the biomass present in immobilized cell preparation, by measurement of the protein or nucleic acid content. These methods are well-known to biochemists and so are frequently used. Carbohydrate content is less frequently used because it is not so easily related to cell biomass nor so easily determined and is also more subject to changes during growth and cell differentiation. Quoting the activity of enzymes in relation to the amount of protein present (specific activity) is frequently used in enzymology when dealing with enzyme extraction and purification. It may be useful when considering immobilized enzyme preparations but its value as a basis for the description of the biocatalytic activity of immobilized cells is doubtful. There will always be a large amount of non-target enzymic and other proteins present. In addition there may be many interfering substances also present which cause difficulties in the protein assay method. In consequence, specific activities are expected to be low in immobilized cell preparations. It should also be remembered that some methods of immobilization involve modifications to the proteins present and this may affect standard protein assays to a significant extent. Other analytical methods based on total organic carbon, organic nitrogen and inorganic element content, such as phosphorus, could also be used with some types of cell support material but have not found general application. A major problem with all the methods based on fresh or dry weight, direct cell counting or biochemical analysis is that no distinction is possible between living and dead cells. For some preparations this may be of no significance because the retention of cell viability may not be of any importance or the cells may even be deliberately rendered non-viable by permeabilization.

The conventional microbiological method of enumerating viable cells is to plate out appropriately diluted samples and either count the number of visible (to the naked eye) colonies developing or estimate statistically on the basis of which dilution gives at least one visible colony developing per plate inoculated (commonly called the Most Probable Number technique). Animal cells can also be enumerated by plating out on an appropriate medium and direct microscopic examination of the plate or a special culture flask which has flat sides. With plant cells there is considerable difficulty in using such methods because a certain number of viable cells must be present before any of the cells will grow and this makes the calculation of the number of viable cells present difficult. Automated colony counters are available for microbial cultures and sophisticated instruments consisting of linked microscopes and computer systems can speed up and extend the scope of these culture-based methods of determining cell viability. Such methods have found little application in studies of immobilized cells because of the difficulty of recovering the cells from the support materials before plating out and culturing.

Because of the difficulties and length of time required many researchers have adopted other methods of assessing cell viability (Table 2.3) that are based upon specific properties of living cells which are known to be reliable indicators of intact cell function. An intact, selectively permeable plasma membrane is essential for viability and many methods have been studied to test for the integrity of this membrane. In cases where the cells themselves can be individually examined with the light microscope a staining procedure which allows the determination of percentage viability is convenient. In simplistic terms cells can be

Table 2.3. *Some commonly used methods of assessing viability of immobilized cells*

Technique	Advantages	Limitations
Increase in cell number	Direct, absolute measure	Often difficult or impossible to observe
Vital staining of living cells	Direct, simple, rapid	May be difficult to observe
		May not relate in a simple way to viability
		Generally needs fluorescence microscope
Staining of dead cells	Direct, simple, rapid	May be difficult to observe
		May not be reliable if empty cell ghosts are present
Ability to reduce redox indicators	Simple, rapid	May not relate in a simple way to viability
Ability to respire	Simple, widely applicable	May not relate in a simple way to viability
Ability to produce ATP	Widely applicable	Relatively complex method needing special instrumentation
		May not relate in a simple way to viability

considered as viable (i.e. having a living protoplast surrounded by a functional plasma membrane), non-viable but still containing the majority of the protoplast contents, or as empty cell-wall ghosts, containing negligible remnants of the protoplast. It will at once be clear that a staining method based on positive uptake by living cells, with an involvement of an enzyme normally functional in living cells would be ideal. Vital stains are those which, in appropriate (generally very low) concentrations will stain living cells. At higher concentrations and/or over long time periods virtually all vital stains are toxic and upset cell metabolism. Many of these react generally with proteins or nucleic acids and so easily distinguish cell ghosts from viable and non-viable cells with contents. Distinction between the latter two categories is more difficult. Fluorescein diacetate is one of the few vital stains which also acts as a substrate for esterases and in living cells is hydrolysed liberating fluorescein, which can be readily detected inside cells with a fluorescence microscope. Toxicity becomes apparent after a relatively short time and the fluorescein leaks out of the cells. The fluorescein diacetate itself also spontaneously hydrolyses and although a stock solution in acetone will keep for long periods the freshly made aqueous stain must be used within about 30 minutes of preparation. No distinction is made between non-viable cells and cell ghosts.

An alternative approach is to use a stain which is positively excluded by living cells. Again, the stain must be of low toxicity and be used at low concentrations. Because the permeability of plasma membranes differs from one species to another there is no one stain which is applicable to all species. Methylene blue is frequently used to distinguish dead yeast cells from living cells which exclude it when exposed to very dilute (of the order of 0.001% w/v) aqueous solutions. Under other situations this stain can act as a redox indicator, changing colour when it is converted from the oxidized to the reduced state, or vice versa. Various derivatives of tetrazolium chloride can also act as redox indicators and in fully or partially functional cells may be reduced by respiratory metabolism. In this use the stain in the oxidized form must be able to enter the cell, where it becomes reduced and the colour change can be seen under the light microscope. Respiratory activity may remain for relatively long periods in non-viable cells. Finally, a distinction must be made between excluded stains which merely penetrate into dead cells or cell ghosts and those which are positively concentrated onto the protein or nucleic acid contents of non-viable cells. The latter category of stain will not be able to distinguish between viable cells which exclude the stain and cell ghosts which do not take the stain up because they have no contents inside the cell walls. An account of the value of immunofluorescence and viability stains (for quality control in yeast cultures) was given by Chilver *et al.* (1978). It will be obvious that direct visual observation of viability is both complicated and not applicable to all types of immobilized cell preparation.

In the search for convenient, generally applicable methods of estimating cell viability attention has turned to the assay of either enzymes, or the products of enzyme reactions, which are closely associated with viability. The use of redox stains has already been mentioned and this assay can be quantified by carrying it out under conditions where the plasma membrane has been deliberately made permeable to the substrate and the products, but without inactivation of the test enzyme system. This allows easy quantification using standard spectrophotometric methods. Because of the relative stability of dehydrogenase systems, however, it is likely that such assays will always over-estimate the amount of viable cells present and careful cross-calibration is needed with a more reliable estimator of cell viability. In the search for short-lived intermediates which are indicators of viability the coenzyme ATP has attracted much attention and forms the basis of standard methods and even of specific instruments designed to measure ATP. The majority of these are based upon the luciferin–luciferase system (generally from fireflies) which liberates visible light in the presence of ATP. Sensitive photomultipliers, such as those in fluorimeters, can be used to quantify the light emitted. Although well-known, the method has not yet been extensively used in the study of immobilized cells, due probably to its lack of general availability in laboratories involved with the study of immobilized cells and the rather expressive assay reagents themselves. Each species of cell may require separate cross-calibration between ATP content and the number of viable cells. A rather similar technique, in which the light emitted during the haematin-catalysed oxidation of luminol to aminophthalate was described by Mason, Pirt & Somerville (1978) during a study of benzene metabolism by polyacrylamide-immobilized *Pseudomonas putida*.

A more convenient approach, which relates more closely to overall respiratory activity, is the assay of changes in oxygen or carbon dioxide concentration by the standard methods. Oxygen electrodes are widely available and convenient to use, allowing accurate measurement of dissolved oxygen. The dissolved oxygen is a function of aeration rate, cellular uptake and the rate of solution degassing so that oxygen electrodes are most conveniently used in closed electrode chambers. Marcipar *et al.* (1979) used a small column reactor, which was saturated with oxygen and then the flow system sealed, to demonstrate a higher oxygen uptake by immobilized *S. cerevisiae* on ceramic supports.

A potentially useful method of measuring oxygen transfer rates was described by Wittler *et al.* (1986). They used a micro-coaxial needle sensor of only $0.7\,\mu$m directly inserted into pellets of *Penicillium chrysogenum*. This could equally well provide significant data about diffusion in immobilized cell preparations. Cellular oxygen uptake rates may vary widely, depending on many factors such as culture environment, nutrient supply and physiological state of the cells. Methods which measure oxygen uptake and carbon dioxide liberation simultaneously are helpful in sorting out some of the complexities of metabolic gas exchange. Gas chromatography with appropriate detectors, gas phase mass spectrometry, infra-red CO_2 analysis with paramagnetic oxygen analysis can all be useful for the analysis of headspace or exit gas. Gilson or Warburg respirometry can be used for the laboratory study of small samples. Membrane-inlet mass spectrometry can be a useful technique for direct measurement of many dissolved substances. The possibilities of this technique have been known for a long time (see Weaver *et al.*, 1978, 1980) but, probably due to lack of general availability of the equipment, have not been extensively adopted despite the fact that it offers, in conjunction with various biosensors, a very simple and widely applicable technique of great potential. For a discussion of the topic of oxygen supply to immobilized cells the reader is referred to Enfors & Mattiasson (1983).

Using Warburg respirometry, for example, Ghommidh, Navarro & Durand (1981) showed that when cells of *Acetobacter aceti* were immobilized from the first phase of a growing culture, onto powdered cordierite, there was an almost immediate increase in oxygen uptake. Cells from the late growth or stationary phase did not show this effect.

Vollbrecht (1980) used gas chromatography to follow oxygen uptake rates by various strictly aerobic bacteria (*Alcaligenes eutrophus, Pseudomonas acidovorans, P. delafieldii* and *Paracoccus denitrificans*) and its relationship to the excretion of metabolites. It was shown that excretion increased with reducing oxygen availability. Wagner & Lang (1979) summarized some examples of the use of various techniques to study immobilized cells.

Doran & Bailey (1986) found that immobilization of *S. cerevisiae* resulted in considerable changes in growth and metabolism. Specific ethanol production was increased as was glucose consumption, but the specific growth rate was reduced with greater internal accumulation of storage and structural polysaccharides. The immobilized cells had a higher chromosome number than the free cells. The system used for immobilization was 4 mm glass beads, grafted with a layer of gelatin which was then treated with glutaraldehyde before loading with cells. A packed-bed reactor was used with a high recycle rate. More research of this calibre is needed to understand the changes that occur when cells are immobilized.

There is little point in presenting a detailed summary of specific applications of the techniques mentioned above. Each situation is different and requires its own criteria for the selection of appropriate methods. It should be noted that while the comparison of free with immobilized cells or the determination of true mass or numbers of viable cells in an immobilized preparation are of interest, it is often sufficient to use a more pragmatic approach. If an immobilized cell preparation contains viable cells it will be capable of showing an increase in biomass when re-incubated in a growth medium. Almost any method may be used to show this increase in biomass, without any need to distinguish between viable and non-viable cells. The only caution needed is to ensure that microbial contamination has not occurred and that the biomass increase is therefore due to a genuine increase in the originally immobilized cells. This may be sufficient for many practical purposes, although interpretation of the relationship between biocatalytic activity and cell biomass may need careful consideration.

Physical characteristics of immobilized cell preparations

Particle size and shape are important characteristics of any immobilized cell preparations. In general, evenly-sized spherical particles have the greatest range of applications to all configurations of reactor and have the most readily predictable properties. Direct observation, under an appropriate magnification, gives a qualitative view of the shape and range of size and should be the first stage of any examination. For quantification, a manual vernier measuring microscope can be used but a

Table 2.4. *Some important characteristics of*
support materials for immobilization

Chemical composition and reactivity
Water regain and content
Particle size
Shape or form of support material
Porosity and pore dimensions
Density
Surface area
Compression behaviour
Resistance to abrasion
Stability under operational conditions

standard computer-linked particle size analyser is obviously preferable
and very much quicker. Irregularly-shaped particles are the most difficult
to characterize and average length/breadth measurements may be
adequate. Generally, particle size and shape is predetermined by the
supplier of support materials as well as the average particle size and
standard deviation or range of sizes. When supports are prepared in the
laboratory and gels are subdivided or prepared as small particles, it is
important to determine the size and shape to assist comparison with other
experimental results. The simplest and often quickest method of particle
sizing, within defined limits, is to use dry or preferably wet sieving.
Standard sieves can be readily purchased and the quantity of support in
each size fraction can be obtained by direct weighing. In general the
Coulter Counter is not applicable to particles of the size used as cell
supports. With particles which can be obtained dry and then swollen to
operational size the relevant measurement is of the swollen particles. In
general, the greater the water regain and swelling the lower the
mechanical strength of the support. With non-porous particles the surface
area, the diameter and the size distribution are of major significance.
With porous particles or gels, the outer surface area may be of little
significance, but the diameter (which will influence diffusional
resistance), together with the size distribution, will be characteristics of
importance. Some important characteristics of support materials are
listed in Table 2.4.

The density of a support material (i.e. its mass divided by its volume) is
of significance in stirred-tank and fluidized-bed reactors and may restrict
the use of inorganic supports to a particular size range or type of reactor.
Volume is generally measured by liquid displacement, for example in a
burette or measuring cylinder using a pre-weighed sample of the support
material. An alternative practical approach is to determine
sedimentation velocity, under defined conditions. This is particularly
useful information in relation to fluidized-bed reactors. For fixed-bed

reactors, the flow resistance, under defined conditions, is an important operational characteristic and will be related to particle size and size distribution, as well as to particle compressibility. The accurate measurement of compression behaviour is difficult and requires an appropriate materials testing instrument, with automatic recording and time–force analysis facilities. Data are usually presented as single particle compression behaviour. Bead fracture occurs at the critical compression force but this will rarely be encountered in actual operation. Deformation is of more significance and has two components. These are reversible (elastic) deformation and irreversible (plastic) deformation. The absolute amount of deformation at a given force can also be determined. Without special instruments, only an estimate of compressibility can be obtained, for example, by visual observation and measurement under defined weight loadings. This is laborious and time-consuming, but of great functional significance for gel beads in fixed-bed reactors. Its practical effects can be determined by the measurement of specific flow resistance. This is measured by manometric methods in the actual column being used. The pressure drop with and without the particles present is determined. Care must be taken to exclude air bubbles completely from the system.

For porous supports and gels the determination of accessible internal volume is of significance. For inorganic supports this may be carried out with standard instruments which measure porosity, generally based on gas adsorption techniques which relate porosity to internal surface area in the particles. Direct observations by electron microscopy, either SEM or TEM as appropriate to the support material, can also be used but may prove more difficult to quantify other than by pore diameter and estimated volume. For non-swelling support materials the internal volume can be assessed by infiltrating hexane or some other relatively non-volatile but low viscosity liquid, under vacuum, into a pre-weighed sample of the dried particles. After briefly blotting to remove excess liquid the particles are reweighed and pore volume calculated. For hydrophilic gels and other wettable supports the internal volume can be determined by infiltration methods. The pre-wetted particles, of known total volume, are immersed in a known amount of a test solution of known concentration. After time for diffusion into the particles, and assuming no significant absorption effects, the reduction in concentration of the test substance is then measured and the appropriate calculation for the increased volume of liquid available to the diffusible test substance is carried out. If continuous readings are taken an indication of the diffusion resistance of the particles or gel can be obtained and related to test substances of particular molecular mass. With gel materials the reverse technique of incorporating a mobile substance into the gel beads at formation and measuring the rate of outward diffusion can also give a useful measure of diffusion resistance. If the pore size itself is needed this can be determined by using macromolecules of known molecular size (such as blue or other modified dextrans) and a reversal of the technique of molecular exclusion chromatography, which is well-known to biochemists as a means of measuring molecular size. Colloidal particles could be used to tackle the problem of pores approaching cellular size

Resistance to abrasion is an important characteristic of any particulate immobilized cell preparations, if they are to be used in a reactor system which requires movement of the particles for efficient operation. This is so for any stirred-tank, air-lift or fluidized-bed reactor. High shear forces arise wherever vessels are vigorously mixed or flow is channelled through confined regions of flow systems. In stirred tanks the Reynolds number provides a single parameter which summarizes the influence of factors of major significance. In particular, the tip speed of the impeller is of major importance for the vessel and the nature of the immobilized cell preparation. For any work on abrasion resistance it is essential to define the conditions of the experiment very clearly and to make sure that the conditions are comparable to those of operational reactors. Buchholz (1979) described a recommended apparatus for laboratory studies. After 50 hours of operation at an appropriate speed the effect of abrasion on the particles can be determined by a variety of means, such as direct microscopic examination, loss in weight, separation and recovery of abraded fragments and so on. The results are generally expressed on a percentage loss basis. In batch vessels the presence of abraded 'fines' may be of little significance but in continuous systems flow may be adversely affected, especially at filters, and outlet streams become contaminated with biocatalyst.

Measurements of the physical characteristics of immobilized cell preparations are important in predicting and explaining their behaviour in various types of reactor and hence in selecting an appropriate type of reactor for a particular immobilized cell preparation and vice versa. It should be noted, however, that prior determination of the parameters discussed here does not necessarily give complete information on the future performance of the particles in an immobilized cell reactor. Particles in bulk may behave differently from individual or small sample behaviour. Where viable cells are present their subsequent growth or metabolic products, such as polymeric substances or liberated gas bubbles, may completely upset the behaviour of the particles in the reactor. The reader is referred to Buchholz (1979) for a more detailed discussion of some of the properties of immobilized biocatalysts.

3 Entrapment, encapsulation and retention by membranes

The methods considered in this chapter span a wide range from the mildest procedures, which retain full cell viability, to some severe techniques where the reagents used during immobilization are definitely harmful to cellular integrity. Although related, in principle, to the natural immobilization of cells in their own extracellular secretions all the methods discussed in this chapter are artificial and do not rely upon particular properties of the cells themselves, or the presence of reactive groups on the cell surface. This is not to say that there are no molecular interactions between the cells and the immobilization system. Undoubtedly, such reactions do occur. It does mean, however, that the methods are generally applicable to all cell types and it is the ability of the cells themselves to withstand the procedures used that firstly influences the choice of method. The operational characteristics of the immobilized cells are also of importance.

Entrapment

The matrix in which cells are entrapped is generally described as a gel. This simple word, however, covers a wide range of different characteristics. At one end are macromolecules held together by relatively weak intermolecular forces, such as hydrogen bonding, or ionic cross-bonding by divalent or multivalent cations. At the other is strong covalent bonding, in which the lattice that the cells are entrapped in can be considered as one vast macromolecule, limited only by the particle size of the immobilized cell preparation. It is inevitable that the matrix in which the cells are entrapped will act as a barrier to diffusion. The practical implications of this depend on the circumstances but, in general, it is found that small molecules, both substrate and product, will be able to diffuse at rates acceptable for any of the gel materials. The actual size and form of the gel material may, of course, cause serious diffusional problems. The exclusion of large molecules may have certain advantages, for example, by preventing harmful degradative enzymes from reaching the cells, but this does not generally appear to be of any great benefit. Indeed, for many potential processes the problem of obtaining adequate access for high molecular weight substrates has been a cause of concern.

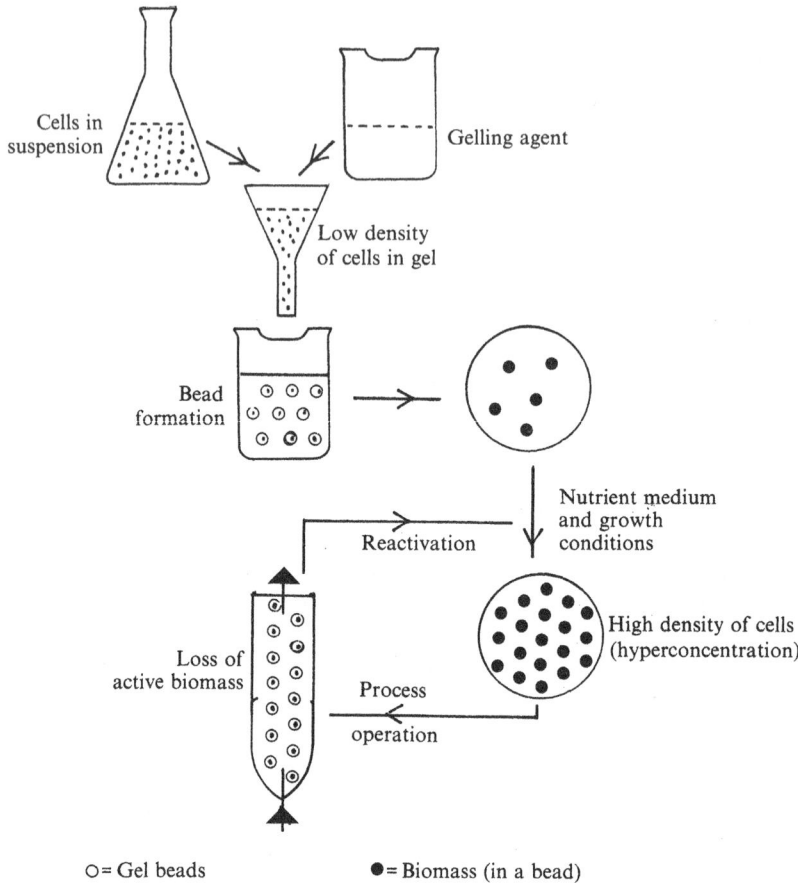

Figure 3.1. Diagrammatic representation of the principle of process intensification by the growth of viable whole cells entrapped in a gel matrix.

It should be remembered, however, that viable cells themselves possess significant permeability barriers – the cell walls and the plasma membranes.

It is at this point, before the actual methods are discussed, that particular attention must be drawn to the two fundamentally distinct types of entrapped cell preparation. Firstly, there are the preparations in which total cell viability (the ability of the cells to increase in size and to undergo nuclear and, where appropriate, cytoplasmic division) is a primary aim. This allows process intensification (Figure 3.1) after immobilization. A few cells loosely entrapped in the matrix may be stimulated into growth by the addition of a suitable culture medium and proliferate to produce very high cell densities. These may be higher, per unit of volume, than can be obtained in free culture in a fermenter. Even where a high cell density was achieved during the immobilization stage, it may be advantageous to reactivate the biocatalyst (i.e. promote the *de novo* synthesis of enzymic proteins), with or without actual cell division. For this type of preparation the cell walls must not have become, coincidentally, in any way fixed permanently to the matrix. Nor must the plasma membranes or any other membrane systems have been damaged so as to alter their inherent selective permeability. Any deliberate attempts at permeabilization are also to be avoided. Gentleness, in every way, is essential for this type of immobilization.

A second type of immobilization involves the entrapment of non-viable cells. The loss of viability may result from a deliberate action on the part of the biotechnologist or be a consequence of the technique and materials selected for entrapment. It may occur after, during or before the actual act of entrapment. Although there may be a loss of certain cell components during the immobilization, the extent of this loss is not important provided the target biocatalytic activity remains as intact as possible. In fact, the activity manifest during practical use may actually increase due to the removal of permeability barriers and the destruction of competing alternative biocatalytic pathways. Loss of viability may be ascribed to a loss of structural integrity by the cell. Compared with the alternative immobilized enzyme preparations there is a cost saving in the absence of the need to pre-extract and then recover the target enzyme. The substances used in entrapment are generally cheaper than the activated supports used to couple covalently with purified enzymes. With fewer unit operations in the overall process there is inevitably less waste to dispose of. The crude cellular preparations may actually show greater stability of the target enzymes due to the protective effect of other proteins. While industrialists often prefer to generate their own viable cell preparations 'on site', there is also already a flourishing trade in non-viable immobilized cell preparations. Trade lists reveal names of best-known producers and the type of biocatalytic activity for which there is a ready market. In an excellent and concise review Gestrelius (1983) lists many other non-viable immobilized cell preparations which have been tested for single and multi-enzyme bioconversions. The great majority, but not all, involved entrapment techniques.

Single-step methods

Table 3.1 lists the major categories of entrapment and the materials normally used for each category. It would be an almost impossible, and pointless, task to list all the various materials that have been used for entrapment. Among authors giving more extended lists and details are Cheetham (1980), Bucke (1983) and Mattaisson (1983). The following selection has been made to emphasize significant points.

Table 3.1. *Some commonly used single-step entrapment methods*

Type	Typical materials used	Limitations
Simple gelation of macromolecules by lowering or raising temperature	Agar, agarose, K-carrageenan, chitosan, gelatin, egg white	Low mechanical strength Possible heat damage
Ionotropic gelatin of macromolecules with di- and multivalent cations	Alginate	Low mechanical strength Breakdown in the presence of chelating agents
Synthetic polymers produced by chemical or photochemical reaction	Epoxy resins, polyacrylamide, polyurethane	Gel precursors are often toxic Some gels have low mechanical strength
Precipitation from an immiscible solvent	Cellulose triacetate, polystyrene	Solvents are often toxic

Agar

Agar is a natural polymeric complex polysaccharide extracted from various species of marine macro-algae. It is readily available from microbiological supply companies and probably the polymer gel most familiar to microbiologists. Several grades are available depending on the extent of purification but it is difficult to obtain information from the suppliers regarding the source and pre-treatments given to the agar. Generally, the purification is intended to make it free from substances inhibitory to the growth of exacting micro-organisms and its use for cell immobilization is very much a fringe market. For immobilization, a dilute solution (generally 2–4% w/v) of the agar is prepared in a medium suitable for the particular cells under study. To dissolve the agar the temperature must be raised well above the setting point of the agar. The temperature is then reduced to a point as close to the setting point as is experimentally convenient. To minimize heat damage to the cells they are not pre-heated to this temperature but simply added, in as concentrated an aqueous suspension as is possible. Usually a 1:1 ratio of cell suspension to agar solution is used. In a comparative study of various matrices to entrap plant cells, Brodelius & Nilsson (1980) mixed at 50 °C. From their results it would appear that some slight heat damage occurred to the plant cells and this possibility has tended to put off many potential users of agar. Special selection of agars setting at lower temperatures is possible, but not particularly attractive to the manufacturers because it might well prove unsuitable for bacteriological plates, which are generally incubated at 37 °C. Various derivatives, such as agarose, are available, although considerably more expensive.

The physical form of the final immobilized cell preparation can be chosen to suit the particular application. It can be cast as a sheet or slab of any desired thickness, which may then be cut up into smaller pieces. A perforated Teflon mould, with holes of 3 mm diameter, was used by Brodelius & Nilsson (1980) to produce short cylindrical beads. Spherical beads can be obtained by adding the molten preparation dropwise to ice-cold buffer. The cumbersome method used by Brodelius & Nilsson (1980) is superseded by the method of pouring into a stirred, heated bath of vegetable oil to produce an aqueous emulsion in oil which is then cooled to 5 °C, with stirring. This technique was developed at the same laboratories (Wikstrom *et al.*, 1982) for a comparable agarose preparation. Banerjee, Chakrabarty & Majumdar (1982) used a much more extreme approach, casting beads containing *Saccharomyces anamensis* (2.5–3 mm diameter) from agar at 45–50 °C into an ice-cold mixture of toluene and chloroform (3 : 1 v/v). After washing with phosphate buffer the beads were air-dried, which increases the mechanical strength of the beads, and the residual β-galactosidase activity measured both then and after storage in a refrigerator. Obviously the preparation may be considered as non-viable. The influence of the concentration of agar used (from 1.5 to 5.0% w/v) was studied. Below 2.5% the matrix was so weak as to allow some cell leakage during the experimental procedure. Above 2.5%, however, the activity of the target enzyme decreased, presumably due to diffusion limitation. At 2.5% agar, nearly 70% of the β-galactosidase activity was retained and the apparent K_m value had risen from 102 mM for the native cells to 148 mM for the immobilized preparation. This apparent lowering of the affinity of the enzyme for its substrate is typical of the immobilized situation and probably indicates diffusional restrictions influencing the transport of substrate and products to and from the active site.

The strength of gels is clearly of great importance for immobilization. Krouwel, Harder & Kossen (1982) presented new comparative results for the tensile strength of slab gels of agar, K-carrageenan, polyacrylamide and calcium alginate using the sophisticated Instron TT/CM instrument. Only the alginate gels showed an inhomogeneous gel structure, with a weaker inner portion and a stronger outer layer. Agar had very poor strain resisting properties. The authors also briefly reviewed compression data for several materials (K-carrageenan, polyacrylamide, polymethacrylamide, calcium alginate and epoxide) but have no data on agar gels. Some workers consider that the major problem with agar is the final mechanical strength rather than the transient high temperatures during preparation (Bucke, 1983). Apart from the plant cells (*Catharanthus roseus*) and yeast species already mentioned, several bacteria (for example *Azotobacter chroococcum*, *Clostridium butyricum*, *Escherichia coli*, *Lactobacillus arabinosus*, *Methylomonas* sp.,

Providencia sp., *Rhodospirillum rubrum*) and other yeasts (*Saccharomyces pastorianus*) have been immobilized in agar, for a variety of biocatalytic studies. There is no theoretical limit to the number of species that could, in theory, be entrapped in agar.

Agarose
Natural agarose is one of the two major fractions found in agar. It is separated, purified and some of the sulphate ester side groups are removed.

It is available in a range of grades, which differ in their physical properties. Low melting point (LMT) agarose is best for entrapment. It is used in the same way as agar, adding the cells to an already prepared solution, as it cools. Gels can then be handled in the same ways. Using what they described as standard low molecular weight agarose, Khachatourians *et al*. (1982) added cells at 50 °C. They used whole cells of *Escherichia coli* and anucleate minicells produced by a mutant with defective cell division. Agarose entrapment is one of the mildest methods available and viability will be retained. The minicells, of course, cannot be described as viable, although they maintain their original structural organization. The activity of the enzyme thymidine phosphorylase was studied using both induced and non-induced cells and minicells. The authors concluded that the greatest specific activity was found with the induced minicells. Some manufacturers supply special grades of agarose which have lower gelling temperatures, Type VII (from Sigma), for example was used by Brodelius & Nilsson (1980) for the immobilization of *Catharanthus roseus*. As expected, the cells retained cellular integrity as demonstrated by respiratory activity and susceptibility to plasmolysis and were capable of growth. They were not considered as active as alginate immobilized cells in secondary product synthesis and biotransformation. Many different species have been immobilized in agarose and almost any could be. One interesting application was to entrap the photosynthetic alga *Chlorella vulgaris* and the blue-green bacterium *Anacystis nidulans*, as reported by Wikstrom *et al*. (1982). The reaction being studied was the oxidative deamination of amino acids. In general, the direct algal amino acid oxidase activity is low but these researchers co-immobilized the *C. vulgaris* with the bacterium *Providencia* sp. PCM 1298, which has a high oxidase activity. The expressed activity rose by a factor of ten, presumably due to the *in situ* production of oxygen by the algae. The oxygen was then available as a reactant for the deamination of leucine.

Kappa-carrageenan
Kappa-carrageenan is a natural polymeric complex polysaccharide extracted from species of marine macro-algae. It has become a popular

gel material for the entrapment of cells. A general account of Kappa-carrageenan immobilization is given by Chibata (1979) whose research group take major credit for introducing this polymer for use in cell entrapment. It is more readily used than agar as a growth substrate by micro-organisms and for this reason, as well as any inherent need for a non-contaminated immobilized cell process, it is often heat-sterilized before use. This is achieved by a normal autoclaving treatment of the stock Kappa-carrageenan solution. Agar, it should be remembered, was selected by microbiologists as an inert growth support gel because it is resistant to degradation by the majority of micro-organisms. Carrageenan is used in the food industry as a thickener and gelling aid and is widely available in a purified 'food grade'. The gelling of Kappa-carrageenan appears to be partly a simple question of temperature. Researchers vary in using temperatures from 50 °C down to 37 °C for mixing of cells and Kappa-carrageenan. It is usually used at an initial stock concentration of 4% (w/v) and often in 'physiological saline' – 0.9% (w/v) sodium chloride in water. Beads could be prepared by the technique of adding dropwise into a cold liquid, as for agar. Pure Kappa-carrageenan gel has very weak mechanical strength and it is virtually essential to ensure that there are plenty of potassium (or various other) ions present to help stabilize the gel. Wang & Hettwer (1982), for example, pumped a *Saccharomyces cerevisiae*–carrageenan mixture into 2% (w/v) potassium chloride solution. This interesting paper addressed itself to solving some of the problems concerning the internal environment within the gel beads. Like so many others, using all manner of entrapment techniques, they found it possible to have very high cell loading inside the beads. Using a similar overall 'inoculum' level they followed viable cell growth and found ten times more cells in a batch reactor holding the immobilized cells than with free cells. Both followed the same typical sigmoid growth curve with similar maximum specific growth rates but the immobilized cells reached the stationary phase plateau at a higher cell density and after about twice the time of the free yeast cells. There was also some cell leakage from the gel beads and growth of these cells in the medium. As expected, supplying more nutrients (in this case yeast extract) resulted in higher cell densities in the gel. Variation in the bead diameter from 3.5 mm to 5.5 mm did not make any significant difference to the final cell concentration in the gel.

Studying the phenomenon in detail Wang & Hettwer (1982) showed no differences in cell mass and mass balance between nutrients for free and immobilized cells. The differences described earlier are merely a result of the mode of expression of the results, because the cells in the gel are clearly present in only a fraction of the total reactor volume and hence process intensification in terms of biocatalytic activity per unit volume has been achieved. Microscopic examination of the gel beads clearly showed the cells in dense microcolonies, with rather more of these towards the outside of the beads. Perhaps the most interesting result was that the inclusion of 5% (w/v) tricalcium phosphate (hydroxyapatite) in the gel gave greatly extended growth of the yeast so that the dramatic drop in viable cells found after about 30 hours in the plain beads did not occur, even after 70 hours. This was attributed to internal pH control within the beads. The crystals of tricalcium phosphate used were 2–30 μm in diameter and exhibited low solubility at neutral pH values. They dissolve at low pH values and hence counteract developing acidity in the gel micro-environment allowing greater cell growth. They can, of course, also be selectively dissolved to any desired extent, from the prepared immobilized preparation by immersion in acidic 1% (w/v) potassium chloride. This naturally results in pores in the gel, where the crystals previously were, and probably helps prevent diffusional restriction of growth. The method is illustrated in Figure 3.2. A further beneficial side-effect of the tricalcium phosphate was to increase the density of the beads. With 10% (w/v) tricalcium phosphate the settling velocity of the beads was doubled. This is a very useful characteristic because it helps prevent bead wash-out at high feed rates in cell reactors. Wang & Hettwer (1982) also suggest the possibility of separating old, less productive beads, from younger ones by the density difference caused by the slow dissolution of the tricalcium phosphate during growth. It should be pointed out clearly that the tricalcium phosphate particles were not here considered as an adsorption support for the yeast, within the Kappa-carrageenan gel matrix, although the possibility of such an interaction cannot be completely discounted.

In other experiments with yeast immobilized in Kappa-carrageenan, Wang *et al.* (1982) showed that cell leakage could be reduced from the beads by increasing the potassium chloride concentration to 4% (w/v) but with a concomitant loss in cell viability. As expected, the more violent the agitation the greater the cell leakage from the beads, but the size of the cells in the beads (comparing normal *Saccharomyces cerevisiae* with a mutant yeast having much larger cells) did not appear to have much influence.

The number of species immobilized in Kappa-carrageenan is very large. Mattiasson (1983), for example, lists 18 different species, covering 11 different products that have been immobilized in Kappa-carrageenan. There would appear to be little restriction upon the type of organism used, ranging from bacteria to fungi and from A to Z in the generic index!

The greater stability of Kappa-carrageenan gels in the presence of potassium chloride was demonstrated experimentally by Krouwel *et al.* (1982), as well as the superiority over agar and inferiority to calcium alginate. Using *Saccharomyces carlsbergensis* Wada, Kato & Chibata (1980) were able to operate a continuous ethanol production system for

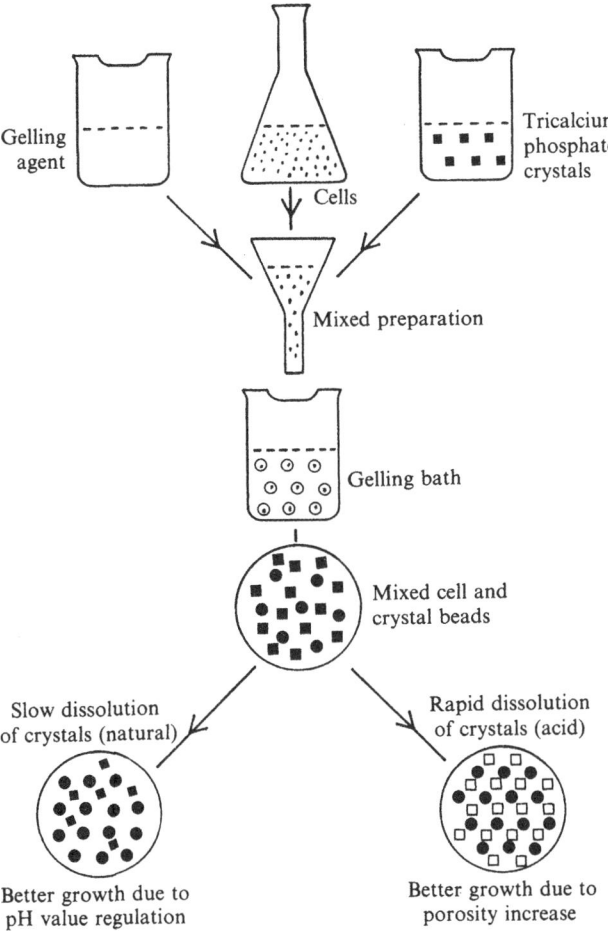

Figure 3.2. Diagrammatic representation of the use of tricalcium phosphate in gel beads.

over three months, with almost 100% of the theoretical conversion yield for the biotransformation of glucose. In their experiments a low density of yeast cells was initially entrapped and the beads then incubated using a complete medium for cell growth. The high cell density resulting from this was maintained over the course of the experiments, with new growth replacing lost cells. The generation time for the yeast of three hours was similar to that of free yeast cells during the initial incubation period. Continuous operation was possible at a retention time of 1.0 hour, over twice the feed rate causing wash-out of free cells. Grote, Lee & Rogers (1980), among many other authors, have investigated the production of ethanol by immobilized bacterial cells of *Zymomonas mobilis*. They compared Kappa-carrageenan with calcium alginate as the matrix. Unlike the results of Wada *et al.* (1980) on *Saccharomyces carlsbergensis* they found a 30% reduction in activity after one month's operation of a continuous process. The *Zymomonas mobilis* was added to 2% (w/v) Kappa-carrageenan at 47–50 °C and used to coat Raschig rings in a rotating flask. These rings were then packed into a column and stabilized with 0.75% (w/v) potassium chloride in 15% (w/v) glucose solution. Over the period of operation of the continuous system bacterial cells continued to be produced and the void space of the column reactor was therefore reduced. Results with the kappa-carrageenan were only slightly superior to those of the calcium alginate entrapped cells.

Considering now the more difficult problem of fungal immobilization, Deo & Gaucher (1983) immobilized conidia of *Penicillium urticae* in Kappa-carrageenan (with added locust bean gum) and allowed them to germinate and grow for 36 hours at 28 °C. The production of the antibiotic patulin was investigated. Immobilization of free conidia which had been germinated for 36 hours at 28 °C before entrapment yielded very poorly active beads and these required a further 48 hours' incubation to bring production up to a level which could be easily measured. Although patulin production in the free cells was typical for that of a secondary product, with very little being produced until the stationary phase was reached, the situation with the immobilized conidia germinated *in situ* was quite different, with significant production during the later stages of the growth phase. The half-life for production was also extended from six days (free cells) to 16 days for the immobilized preparation. The conclusion drawn was that the immobilized cells exhibit a physiological heterogeneity, probably due to nutrient supply. This may well influence the pathways of primary and secondary metabolism in different ways.

Kim *et al.* (1982) considered the biotransformation of progesterone by *Aspergillus phoenicis* immobilized in various matrices. In addition to Kappa-carrageenan and calcium alginate gel entrapment they also used urethane polymers and cross-linking with albumin and gelatin foams. The two cross-linked preparations gave very poor conversion. Calcium alginate gave the greatest overall conversion but much of this was to 15-β-hydroxyprogesterone and especially 6-β-hydroxyprogesterone. Kappa-carrageenan and polyurethane both gave greater conversion to 11-α-hydroxyprogesterone, the precursor desired for cortisone production. This is one of the few papers concerned with the more subtle aspects of the influence of immobilization on secondary product formation, a topic which will clearly develop rapidly.

In one of the less common studies involving the production of extracellular enzymes by immobilized fungi Frein, Montenecourt & Eveleigh (1982) used *Trichoderma viride*. They used a mycelial homogenate prepared from a five day culture grown on 1% Avicel 105 (a particulate cellulose preparation). This drastic disruption process, coupled with the fact that they added it to Kappa-carrageenan at 50 °C cannot have helped to retain metabolic integrity, but they still managed to exceed the cellulase production of a single-stage continuous culture system using free mycelium and approached those of a two-stage continuous system. The specific activity with respect to cellulase activity was also high, implying an efficient conversion of substrate into target enzyme protein. It can be reasonably speculated that they may well have been even more successful using the methods of Deo & Gaucher (1983). In their study of *Catharanthus roseus* cells immobilized in Kappa-carrageenan (mixed at 50 °C), Brodelius & Nilsson (1980) also found some evidence of heat damage. The bioconversion of tryptamine to ajmalicine was less than in free cells or those immobilized in agarose or calcium alginate.

Proteins
Collagen is an animal protein derived from connective tissues such as skins and cartilage. It can be dissolved at low pH values and recovered from solution at higher values. It has been widely used as a support for enzymes with the final product being cast into membrane form. To stabilize the structure it is usual to tan the protein using glutaraldehyde, which cross-links the protein molecules. Most protein methods for whole-cell immobilization also use this reaction to stabilize the matrix. Although the initial coagulation of collagen with the entrapped cells has been generally considered to be largely due to hydrogen bonding forces, there does appear to be a possible involvement of the lysine residues of the collagen in an essentially adsorption process (Cheetham, 1980). Use of glutaraldehyde, however, makes this of no significance because it introduces extensive covalent bonding and consequent high stability of the immobilization matrix (Figure 3.3). Excessive exposure to glutaraldehyde will certainly cause damage to cell function and some care needs to be exercised in getting the conditions right for a particular method of preparation. Cells immobilized with a process involving glutaraldehyde are generally not capable of regrowth and division. Vieth & Venkatsubramanian (1979) give a brief but interesting overview of their patented system involving a membrane cast from tanned collagen and wound with an inert plastic spacer material. They list eight bacteria, *Aspergillus niger*, mammalian erythrocytes and chloroplasts as being successfully immobilized with collagen. Their list has been extended by other workers, although interest in this method does not currently appear to be particularly great. This is most probably due to the superior quality of another protein preparation from the same source – gelatin. This is available, cheaply, as a food grade material.

Tramper *et al*. (1979) entrapped *Arthrobacter* strain X-4 to provide a

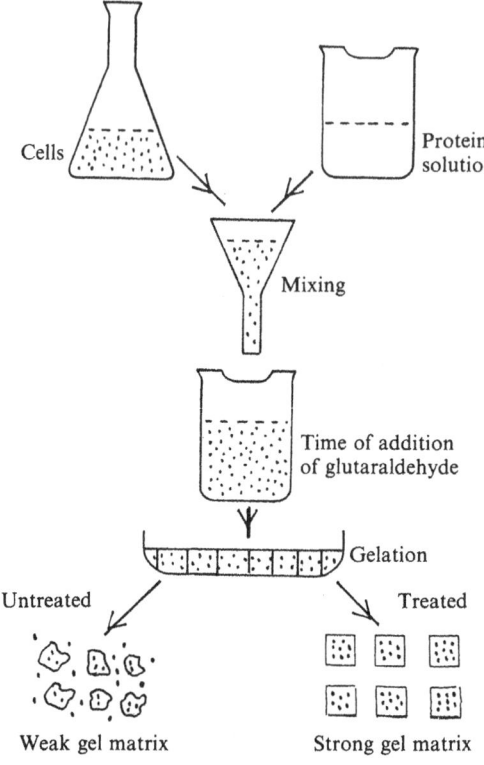

Figure 3.3. Diagrammatic representation of the principle of glutaraldehyde cross-linking of protein gels.

preparation with a high xanthine oxidase activity. The cells were pre-adapted to xanthine before immobilization. These cells were then suspended in a gelatin solution and freeze-dried with glutaraldehyde cross-linking. The resulting material was then milled to three different particle sizes (0.5, 0.7 and 1.0 mm nominal diameter). Although a range of concentrations of gelatin was used, most experiments were carried out with a 1:1 ratio of cells to gelatin (on a dry weight basis) with a 10% (w/v) initial gelatin solution. This was easy to handle at all stages, with acceptable mechanical properties in the final particles. Only very small-scale columns were used in this study but the author concluded that their cell preparation had potential as a continuous biotransformation system. At 17 °C the half-life of the xanthine oxidase activity was about eight and a half days.

The invertase activity of *Saccharomyces cerevisiae* was investigated by Parascandola & Scardi (1981). These immobilized yeast cells cannot ferment glucose or sucrose but do show a more stable invertase activity than free cells. Their freeze-dried yeast was entrapped at a 1 : 10 (dry weight) ratio with gelatin. Compared to free cells the pH optimum was scarcely affected, although the invertase activity of the immobilized preparation fell off more rapidly at pH values above the optimum (pH 4.65). Apparent K_m values for batch hydrolysis were 40 mM for free whole cells and 50 mM for the entrapped ones. On the assumption of ideal plug flow in a tested column reactor, however, the immobilized cells had an apparent K_m of 36.4 mM. Free invertase, of course, has a lower K_m value.

In their studies with *Catharanthus roseus* Brodelius & Nilsson (1980) used gelatin and both a gelatin-agarose and a gelatin-alginate copolymer all with glutaraldehyde cross-linking. As might be expected, cell growth and respiration were adversely affected by all methods involving glutaraldehyde treatment. The structural integrity of the plasma membrane was, however, retained but it may be assumed that its role as a selectively permeable membrane was adversely affected. These authors therefore reject any method involving glutaraldehyde for plant cells. This does not mean, of course, that the more resilient enzymatic activities found in plant cells are necessarily also destroyed. Applications could be developed for these, but it is generally easier to find a microbial source of this simpler type of biocatalytic activity.

In theory, almost any readily available protein can be used for entrapment. When glutaraldehyde is used as a cross-linking agent it is not even necessary for the protein to form a gel as a result of its own inherent properties. There is a plentiful supply of unpurified hen egg white and this has been studied by De Rosa, Gambacorta, Lama & Nicolaus (1981). They worked with the thermoacidophile archaebacterium *Calderiella acidophila* which is capable of growth at 87 °C and pH 3.0. Fresh cells were mixed with liquid egg white at 0 °C and glutaraldehyde in phosphate buffer was stirred in. The mixture was then vacuum rotary evaporated (60 °C), ground and washed. A similar method was used to prepare a magnetic matrix by the inclusion of magnetite particles of 0.3–0.7 μm diameter. The observed activity of β-galactosidase was approximately thirty times greater than that of comparable free cells. At 70 °C over 50% of the activity was stable for up to 200 hours. The magnetite did not appear to cause any adverse effects on the properties of the preparation (Figure 3.4).

Hen egg white has the interesting property of containing large quantities of lysozyme, an enzyme capable of lysing many species of bacteria. It could therefore be of interest as a continuously self-sterilizing entrapment matrix, overcoming some of the problems associated with maintaining sterile operating conditions over extended periods in continuous processes. Kaul, D'Souza & Nadkarni (1983) investigated this possibility using *Escherichia coli* as the test immobilized cells and *Micrococcus lysodeikticus* as the challenging bacterial contaminant. The

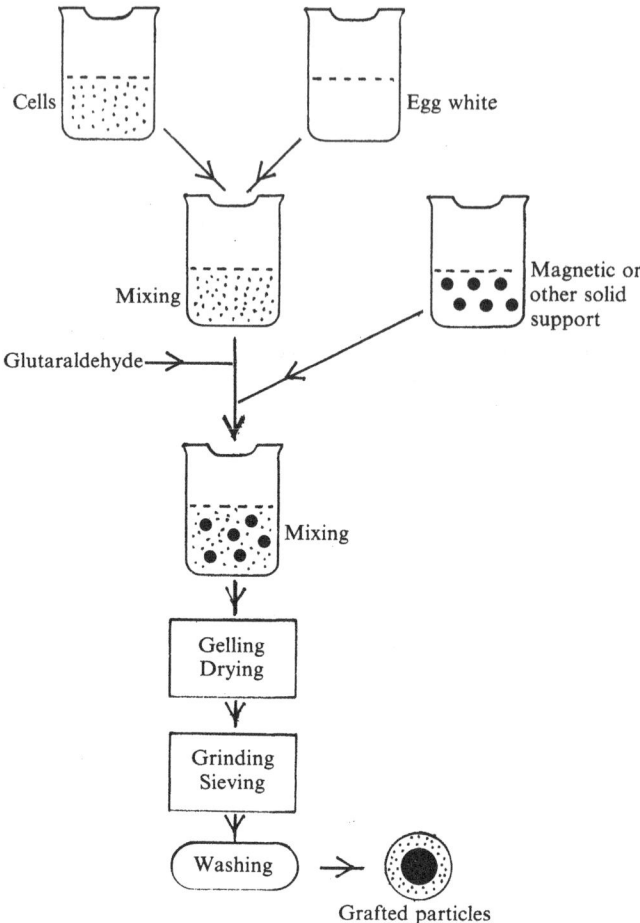

Figure 3.4. Diagrammatic representation of the principle of grafting an entrapment gel matrix onto an inner solid or magnetic support material.

egg white was obtained from fresh eggs and glutaraldehyde used as the cross-linking agent. The control used polyacrylamide gel entrapment. An approximately constant 30% cell lysis was observed for the seven days of study, in a packed bed reactor, using the egg white-immobilized preparation. None was observed in the control reactor. Obviously, the exact concentration of glutaraldehyde and the conditions used affect the amount of observable lysozyme activity but it was concluded that sufficient remained, under conditions giving a satisfactory immobilized cell preparation, to be of benefit. The method is certainly cheaper than the co-immobilization of cells with purified lysozyme, as carried out by Mattiasson (1979).

Alginates

Perhaps the most studied method of entrapment is the ionotropic gelation of alginic acid by multivalent cations, in particular by calcium ions. Alginic acid is a heteropolymer of L-guluronic acid and D-mannuronic acid but the actual proportions of the two vary with the source. The alginates are extracted from several different species of marine algae and, after processing, are available as water-soluble sodium salts in a wide range of grades. Many purified 'food grades' are available. The gel strength is greatest in preparations with high L-guluronic acid contents, such as those drived from species of *Laminaria*. High D-mannuronic contents are found in alginates from *Ascophyllum* and *Macrocystis* and this tends to give a larger internal pore size to the gel beads. Ionotropic gelation is very easy to carry out. Cells are mixed with sodium alginate solution and added dropwise to dilute aqueous calcium chloride. This mild method retains cell viability but is offset by the ease with which chelating against, such as phosphate and citrate buffers, can disrupt the beads. The rather weak gel also allows cell leakage, particularly where growth is continuing after entrapment. Stirred or shaken beads are liable to mechanical damage. One simple way of improving stability is to dry the beads, which causes an essentially irreversible shrinkage. Krouwel *et al.* (1982) showed that calcium alginate gels are not homogeneous and, even before drying, have a stronger outer layer. Drying can result in 100-fold increases in compression strength and clearly the size of beads will exert an important influence on mechanical properties due to the consequent difference in surface area to volume ratios. As the concentration of alginate in the beads increases so will mechanical strength but increasing the cell mass present will obviously have the reverse effect.

Ordinary laboratory grade sodium alginates (or alginic acid, sodium salt as some suppliers describe it) generally shows considerable heterogeneity with respect to the length of the polymer chains present, resulting in considerable differences in gelling characteristics. The careful researcher will take great care to obtain specifications for his alginate if reproducible results are required from one batch to another. Differences between suppliers will often be highly significant with, for example, a 1% (w/v) gel of one being of a comparable gel strength to a 4% (w/v) gel of another. The simple model system, measuring the diffusion of NAD and haemoglobin in alginate gels, under a variety of conditions, used by Kierstan, Darcy & Reilly (1982) provided useful base-line information. For simplicity of study they used fibres of alginate gels rather than beads or slabs. Under otherwise constant conditions, the concentration of calcium ions (range 0.125–0.5 M) had no effect on protein diffusion through 1% (w/v) gels. As expected, diffusion of large, but not small, molecules is hindered by increasing alginate concentrations up to 4% (w/v).

In studies on the diffusion coefficients of glucose and ethanol in alginate membranes, Hannoun & Stephanopoulos (1986) showed that 2% alginate was comparable to water but that the diffusivities decreased with increasing concentrations of alginate. The hydrodynamic theory of diffusion fitted the data well and would probably be applicable to other substrates and products. In an interesting comparison of cross-linking, entrapment (alginate, K-carrageenan and polyacrylamide) and attachment onto titanium (IV) hydrous oxides, Willetts (1986) showed that different methods of immobilization gave the best production of butane-2,3-diol from starch and from glucose. The organism used was *Aeromonas hydrophila*. It was concluded, as might be expected, that diffusional limitation in the gels made entrapment unsuitable for the use of starch as a substrate.

In their work with alginate-immobilized *Chlorella*, Dainty *et al.* (1986) showed the importance of providing environmental conditions which stabilize gel structure and minimize growth if prolonged use of alginate beads is contemplated. The beneficial effect on gel strength of using trivalent cations was demonstrated by Rochefort, Rehg & Chau (1986). They found a two-fold increase using 0.1 M aluminium nitrate as a treatment for calcium alginate beads.

The problem of scaling up manual methods for the production of gel-entrapped cells has been tackled by Matulovic, Rasch & Wagner (1986) who describe a rotating nozzle-ring spraying the cells, mixed with the gelling agent, into a rotating vessel of the cross-linking solution. Large quantities of controlled-sized beads of alginate or carrageenan were produced and this type of apparatus has potential for a wide range of other gelling materials. An alternative approach to the large-scale production of alginate beads was taken by Rehg, Dorger & Chau (1986). They used a dual fluid atomizer in which sodium alginate beads were sheared off the tips of hypodermic needles by an air stream into calcium chloride solutions. The size of the beads was of the order of 1 mm diameter and consistent with theory in relation to the effect of fluid viscosity and air-flow rate.

Because of its mildness, the alginate method can be used to immobilize any type of cell, with the retention of maximum biocatalytic flexibility. The method is simple and safe under any laboratory conditions and has been evaluated many times with bacteria, yeasts, fungi and higher plant cells. Only a few selected results will be considered here, taken from the very extensive literature.

The production of ethanol by both bacteria and yeasts has been particularly well studied. Grote, Lee & Rogers (1980) used a specially selected strain of *Zymomonas mobilis*. They used a syringe to inject a suspension of cells, at the late exponential stage, into the gaps between perforated plates in a column reactor, filled with a 0.75% (w/v) calcium chloride, forming what they describe as fibre-like masses. Up to a dilution rate of 0.3, glucose (at 15% w/v) was completely taken up. Sufficient new cell growth occurred to more than replace lost cells. Maximum production of ethanol was obtained with a dilution rate of 0.85 although both cells and unused glucose were presented in the outflow. They showed continued ethanol production for 800 hours, with around 30% loss in productivity when expressed per litre of outflow, although yield on a gram per gram basis remained constant. They made some comparison with the same bacteria trapped in K-carrageenan but the interpretation is difficult because in the latter case the gel was coated onto Rasching rings rather than formed by injection. In this case void space in the column became partly filled with increasing biomass. The authors calculated a higher maximum productivity for the K-carrageenan gel. As expected, the specific rates of glucose uptake and ethanol production are less for the immobilized cells than for cells in free suspension, by a factor of about 50%. This has to be set against the greater ease of operation with the immobilized cell reactors.

Margaritis, Bajpai & Wallace (1981) used a more conventional alginate bead system (1 mm diameter) with Z. *mobilis* and with a very high cell density (5.8%, dry wt/bead volume). They emphasized in their paper the essential need to define clearly the exact nature of the calculation of results for ethanol productivity to enable proper comparisons to be made. They used dilution rates up to 3.9 and obtained maximum ethanol productivities of over twice that of many other workers with whom they compared their results. The reactor was operated continuously up to 168 hours and the feed glucose stream did not contain calcium ions, resulting in a steady decline in ethanol productivity. A short treatment (15 minutes at that time) with 4% (w/v) calcium chloride boosted the activity to levels above that at 12 hours, although it took 48 hours from the time of the calcium chloride treatment to reach this point and a second decline then set in. Their feed to the reactor, which was set up aseptically to avoid contamination, was 10% (w/v) glucose in a medium of yeast extract plus inorganic salts. In addition to its role in stabilizing the alginate beads, they pointed to the essential role of calcium ions in cellular metabolism. From the known involvement of calcium in the functioning of cell membrane permeability systems, it is likely that this is of significance in the interpretation of their results. As is so often the case, the reactor used for these studies was of a special design. It allowed the escape of the gas produced during fermentation and is discussed later (Chapter 7).

The normal relationship between gel and liquid in a reactor was reversed by Johansen & Flink (1986). They arranged the system so that calcium alginate was gelled directly in a reactor around a regular pattern of rods. When the rods were removed a gel block with internal flow channels was formed. The system was particularly suitable for immobilized cell preparations which produced gas and showed good stability over several months. The test system was ethanol production by *S. cerevisiae*.

A series of three inverted conical reactors was used by Klein & Kressdorf (1983) in their studies of *Z. mobilis*. They used a controllable bead-forming apparatus which allowed easy production of beads of any given size between 0.5 and 3.0 mm. In the experiments reported, beads of 0.8 mm average diameter were chosen for their good diffusional characteristics. Although a low cell density was initially immobilized, the cell density was deliberately increased by five days of continuous (dilution rate 4.0) nutrient feeding. Comparative tables produced by Klein & Kressdorf, including the results of other researchers, again demonstrate enhanced productivity and it is likely that this trend to improvement will continue as new operational systems and *Z. mobilis* strains are developed.

Comparative studies of alginate immobilized *S. cerevisiae*, *S. uvarum* and *Z. mobilis* were carried out by McGhee *et al.* (1982), using their own novel design of reactor. Their paper is also notable for the attention given to the determination of cell viability. Using the SEM, most of the cells were shown to be at the surface of the beads. The beads were made of 1% (w/v) alginate and, when necessary, were disrupted in 10% (w/v) sodium tripolyphosphate for viable plate counting. They concluded that *S. cerevisiae* was the most productive species. After eight days operation approximately 30% of cells were viable, falling to 3% after 16 days. For *S. uvarum* and *Z. mobilis* viable figures of 10% and 12% respectively after 12 days are quoted. Unfortunately, the more detailed results on viability are not given. The multiple discs mounted on a central shaft in the column reactor are described as eliminating the channelling of the glucose feed and the heterogeneous distribution of the alginate beads. They gave an approximately three-fold increase in ethanol production for the yeasts but had little effect on the *Z. mobilis*, under continuous operation.

Aseptically recycling alginate-immobilized cells in a batch fermenter, Margaritis & Bajpai (1981) obtained 70% of the initial activity after 11 cycles. They used *Kluyveromyces fragilis* for the production of ethanol from extract liquors of the Jerusalem artichoke. The yields were approximately 96% of theoretical. The beads (2 mm approximate diameter) were immersed in 4% (w/v) calcium chloride between batches. Each batch was run at 35 °C for seven to nine hours. Haggstrom & Molin (1980) studied acetone–butanol–ethanol production by *Clostridium acetobutylicum*. They used both vegetative cells and spores immobilized in alginate. The spores were heat-activated (95 °C for 30 minutes) in the gel and allowed to germinate before washing in a non-growth medium. Anaerobic conditions were maintained by using nitrogen gas for flushing. The rate of butanol production fell rapidly over the first few days, when vegetative cells in the solvent production stage of batch growth were used for immobilization. This was not seen with spores germinated inside the alginate beads. The method can be expected to be a method of general application to other organisms and types of immobilization. Using a continuous stirred-tank fermenter (CSTR) Krouwel, Groot & Kossen (1983) investigated the isopropanol–butanol–ethanol fermentation with *Clostridium beijerinckii*. The immobilized cells could be reused several times although individual fermentations showed sudden and rapid losses of activity after 15–27 days. Spores were present in the beads and formed the starter for subsequent runs with the same beads. The CSTR gave greater productivity than a continuous column fermentation and a free-cell batch fermentation. These authors also present a mathematical model for steady-state conditions of this fermentation.

The conversion of substrate carbon into fermentation products is particularly efficient in homofermentative lactic acid bacteria and Stenroos, Linko & Linko (1982) have investigated the properties of alginate-immobilized *Lactobacillus delbrueckii*. A simple glass, packed-bed column was used for continuous and batch recycle runs. In the latter case the substrate (glucose 4.8% (w/v), yeast extract 1% (w/v) and 4.8% (w/v) powdered calcium carbonate acting as reserve buffer) was circulated continuously. Yields of up to 97% of theoretical lactic acid yield were obtained in 40 hours at 43 °C. Of particular interest was the fact that the immobilized cell preparation could be stored at +7 °C, in 4.8% (w/v) glucose, for at least 55 days with only a slight reduction in activity at first batch reuse. Regrowth occurred both before and after storage so that a roughly constant activity was maintained with as little as 10% loss of activity over 32 separate batch reuses, extending over 157 days. This draws attention to the particular value of using viable immobilized cells as a self-regenerating biocatalyst system. As might be expected, however, long residence times (10–20 hours) are needed for maximum yields (as percentage of theoretical) in continuous mode, due to excessive growth at higher dilution rates. The half-life was around 100 days with an approximately 12 hour retention time.

The immobilization of fungal hyphae presents problems. One solution is to grow the fungus as pellets. Matteau & Saddler (1982a) used this for *Trichoderma viride*. Pellets of approximately 2 mm were immobilized in alginate and the activity of β-glucosidase investigated using cellobiose and salicin as substrates. An upflow packed column reactor was used. With both salicin and cellobiose there was evidence of product inhibition, shown by cyclical changes in effluent glucose concentrations. The cell-associated β-glucosidase activity had a half-life of just over 1000 hours at 50 °C, based on 340 hours of continuous operation. Livernoche *et al.* (1981) used a glass ball (2 cm diameter) inside flasks of a culture of a fungus *Coriolus versicolor* on an orbital shaker. This ensured that the mycelium remained in suspension as fine fragments, which could easily be incorporated into alginate gel beads. These were active in decolorizing Kraft mill effluent, probably by degrading the oxidized and chlorinated lignins present in the effluent. Batch flask experiments were carried out over four days. A later paper (Royer *et al.*, 1983) gives details of the biocatalytic activity in an air-lift reactor. It might be expected that hyphal outgrowth would occur from beads containing viable mycelial fragments unless adverse nutrient or environmental conditions exist in the immobilized cell reactor.

The cells from liquid suspension cultures of higher plant cells are frequently immobilized in alginate, due to the simplicity and gentleness of the method. Most culture media contain sufficient quantities of phosphate to weaken the beads, particularly if cell growth is also occurring. Jones & Veliky (1981) developed a maintenance medium lacking plant hormones, vitamins and ammonium ions and substituting 2-(*N*-morpholin)ethanesulphonic acid for phosphate. By varying the sucrose and mineral element content a medium retaining viability for at least 24 days was developed. The cells (of *Daucus carota*) also retained the ability to biotransform digitoxigenin. Viability was assessed by the ability to take up oxygen, using an oxygen electrode. Replacement of the maintenance medium by a growth medium gave demonstrable new growth. This would allow a controllable reactivation of immobilized plant cell preparations that are showing progressive loss of biocatalytic activity. With plant cells, the use of a syringe is not recommended for preparing the alginate beads, because it causes shear damage to large cells and clumps of cells. The formation of clumps of cells is a disadvantage of the conventional suspension growth of many plant cell lines. An interesting use of alginate-immobilized cells is to produce suspension cultures consisting of mostly single cells. This is done by allowing regrowth from beads containing viable cells. The process can be repeated many times. Morris & Fowler (1981) describe the successful use of this method for *D. carota, Nicotiana tabacum* and *Catharanthus roseus*. The subject of immobilized plant cells was briefly overviewed by Brodelius & Mosbach (1982).

Using alginate-immobilized *Mycobacterium*, Brink & Tramper (1986a,b) showed that internal pore diffusion of the limiting substrate was the main mass-transfer resistance. A bubble column reactor was used for these experiments and the epoxidation of propene was the target reaction. The results found are those to be expected with such a combination.

By means of theoretical calculations and experimental data, Adlercreutz (1986) was able to show that *p*-benzoquinone was an effective electron acceptor and could sustain higher oxidation rate of glycerol to dihydroxyacetone than could oxygen. This was related to the solubility differences and a reduction in internal mass transfer limitation in the alginate beads used to immobilize *Gluconobacter oxydans*. Perfluorocarbon as used, for example, by Damiano & Wang (1986) may prove to be a solution to oxygen limitation in some immobilized cell preparations. The use of such compounds in artificial blood is well-known. An interesting influence of alginate immobilization was reported by Grizeau & Navarro (1986). They found that immobilization promoted the extracellular excretion of glycerol by the marine alga *Dunaliella tertiolecta*. This particular alga does not produce a cell wall and must be kept in saline conditions but the presence of 20 mM calcium chloride in the culture medium was sufficient to stabilize the beads. This type of wall-less alga can be considered as equivalent to a protoplast in terms of its handling during immobilization.

Chitosan

An alternative approach to the use of polyanions, such as alginate, is to use polycations and cross-link with multivalent anions. This avoids the serious problems caused by chelation of calcium by phosphate, found when using alginate. Vorlop & Klein (1981) used two types of chitosan and the following polyanions: polyphosphates, poly-(aldehydo-carbonic acid) and poly-(1-hydroxy-1-sulphonate-propen-2). The tryptophane synthetase activity of *E. coli* was investigated. A few grams of chitosan (the actual amount depending on the type used) are dissolved in dilute acetic acid, giving a final pH value of around 5. The cells are mixed into the viscous chitosan acetate solution and at once added dropwise to 1.5% (w/v) sodium polyphosphate at pH 5.5. After 30 minutes the beads are transferred to polyphosphate at pH 8.5, which shrinks the beads. The influence of pH value is rather complex since under alkaline conditions the chitosan is precipitated out and the beads are stable in the presence of most ions. At acid pH values the polyanions, promoting ionotropic gelatin, are essential. Further shrinkage and hardening can be carried out by air-drying. The resulting beads have very similar mechanical stability to dried alginate beads but have the superior stability in the presence of phosphates. As with alginate, the viability of cells may be retained and regrowth is possible in chitosan beads. Additional stability of chitosan beads can be obtained by glutaraldehyde treatment.

Further work in the same laboratories (Kluge, Klein & Wagner, 1982) extended the method to the fungus *Pleurotus ostreatus*. Mycelial pellets were disrupted in dilute sodium chloride, washed and immobilized in chitosan. Of the polyanions tested for gelation, polyphosphate used alone gave greater retention of enzyme activity than hexacyanoferrate or polyphosphate followed by glutaraldehyde cross-linking. The reaction studied was the conversion of penicillin *V* to 6-aminopenicillanic acid. With polyphosphate the half-life was extended 10-fold over that of free cells. Relative activity was 38% of free cells. A further application of the technique was described by Stocklein, Eisgruber & Schmidt (1983) who studied the conversion of phenylalanine to tyrosine by a *Pseudomonas* sp. They confirmed the high stability of chitosan beads over alginate, although this was attributed mainly to shrinkage. The former beads were only half the diameter of the latter. The chitosan beads dissolve in the absence of phosphate and swell in acid conditions, but are stable in the presence of chelating agents and monovalent cations. Cell leakage was much less from chitosan beads, but the retained enzyme activity was greater in the alginate beads. Unlike most reports on the activity of immobilized cells these authors discuss interesting modifications of the reaction medium to optimize bioconversion. The conversion requires two coenzymes and a system to regenerate them.

Polyacrylamide

Polyacrylamides are synthetic polymers which can be produced with a wide range of properties depending on the reagents and conditions used during preparation. They are familiar gels in biochemical research, being used for the separation of molecules, chiefly on the basis of molecular size and shape. The work of Mosbach & Mosbach (1966) is generally considered to be one of the earliest examples of cell immobilization by entrapment. They demonstrated retention of the orsellinic acid decarboxylase activity in entrapped, powdered, thalli of *Umbilicaria pustulata*. This is a lichen and also represents one of the earliest examples of the deliberate co-immobilization by researchers of two symbiotic species. Nature, of course, developed it millions of years ago! The development of microbial cell immobilization and many applications are reviewed by Cheetham (1980).

In the preparation of gels a monomer is reacted with a bifunctional cross-linking agent. The monomer is usually acrylamide, although methylacrylamide is sometimes also used. The bifunctional reagent is usually N,N'-methylene-bis-acrylamide (BIS for short), which is readily available. Chibata & Tosa (1976) compared seven bifunctional agents for the immobilization of *E. coli*. Only N,N'-diallyl tartardiamide gave better (slightly) results than BIS, in terms of yield of aspartase activity. 1,2-diacrylamide ethyleneglycol, N,N'-propylenebisacrylamide and diacrylamide dimethylether were slightly worse than BIS. The activities with ethylene urea bisacrylamide and 1,3,5-triacryloyl hexahydro-S-triazine were ten times less. The mechanical strength of a gel is stated to increase in proportion to the square root of the concentration of acrylamide monomer used with a given fixed concentration of cross-linking agent. At the same time, however, the pore size of the gel decreases, thereby causing diffusional restrictions to cellular activity. Larsson, Ohison & Mosbach (1979) achieved least cell damage by using relatively high concentrations of tetra-methylethylenediamine and ammonium persulphate (used as initiators of polymerization). With *Arthrobacter simplex*, they suggested that polymerization should occur within 60 seconds after mixing. A rapid rate of reaction, however, leads inevitably to a rapid temperature rise and appropriate physical arrangements must be made to dissipate the heat of reaction generated during polymerization. Polymerization is always carried out in the presence of a suitable buffer. The adverse diffusional properties of gels with small pore sizes can be partly overcome by incorporating a high concentration of cells in the mixture, thereby gaining the advantages of the higher mechanical strength of small pore gels. Obviously there is an upper limit to the concentration of cells that can be used. It is possible to initiate polymerization with ultraviolet light if riboflavin (Mattiasson, 1983) is used in place of ammonium persulphate.

One method which may reduce the cellular damage caused during entrapment is to mix the cells with preformed polymers in place of the monomers, leaving only the final cross-linking to be completed. The polymers may be modified, for example with acylhydrazide groupings, to confer water solubility and allow mild entrapment conditions, if the appropriate cross-linking agent (in this case dialdehyde) is added in carefully controlled amounts (Freeman & Aharonowitz, 1981). Another modification of the basic method is covered by Japanese Patent no. 138414 in which acrylic acid and its salts are mixed with the acrylamide monomer before polymerization. When mild methods are used, regrowth is possible in polyacrylamide gels.

In the simplest methods, the polyacrylamide gels are cast into slabs or blocks and mechanically converted into small particles after polymerization (Figure 3.5). This is frequently a rather uncontrolled stage and may result in particles of irregular or flat-faced shape, which are less desirable than spherical beads in many types of reactor. The Patent Applications of Rosevear (1982) include, amongst other systems, the use of thin sheets of material coated, before polymerization, with a viscous polyacrylamide-generating/cell system. When gelled, these sheets can be rolled for use in a tubular reactor. It is possible to produce regular-sized beads of polyacrylamide, as described by Kostner & Mandel in Mosbach (1976) but this does not appear to have been generally adopted.

The mechanical properties of polyacrylamide gels, compared to alginate, agar and K-carrageenan, were investigated using the Instron TT/CM apparatus by Krouwel, Harder & Kossen (1982). Unfortunately, they do not specify the exact conditions of polymerization and so the results for polyacrylamide are difficult to interpret. Klein *et al.* (1978) obtained maximum mechanical strength with a gel produced from a mixture of prepolymerized polyacrylamide in a suspension polymerization of methacrylamide in dibutyl phthalate, although this was

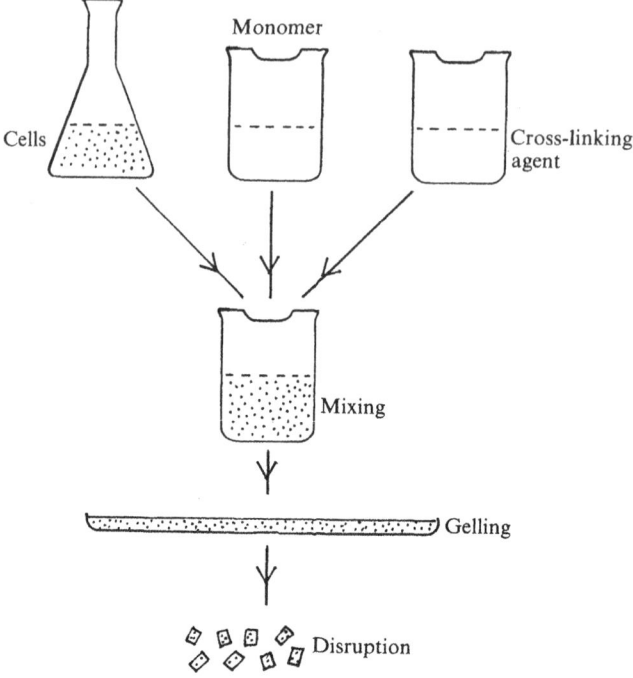

Figure 3.5. Diagrammatic representation of the principle of polyacrylamide gel formation.

at the expense of yield of activity, which was only half of that using a simple polyacrylamide gel system. This most brief paper gives reference to researches reported in two theses. One of the most thorough studies of the diffusional properties of polyacrylamine gels containing immobilized *Alcaligenes faecalis* is given by Wheatley & Phillips (1983a). Their polymerization system was a standard acrylamide plus BIS with β-dimethylaminopropionitrile as promoter. A variety of particle sizes was obtained by cutting, extrusion through meshes and blending. The activity of β-glucosidase was measured chromogenically with paranitrophenyl-β-D-glucoside. Both a stirred reactor and an upflow packed column were used. Apparent V_{max} and K_m values were obtained. As expected, smaller particles had higher V_{max} values, indicating internal diffusional restriction in the larger particles. Of importance to the interpretation of any results on activities in immobilized systems was the finding that V_{max} values obtained by extrapolation from high external substrate concentrations were greater than those when using low external substrate concentrations. This emphasizes that such data should be obtained under realistic operational conditions for satisfactory prediction of process efficiency. In a further paper, Wheatley & Phillips (1983b) described the influence of polymerization conditions upon the retention of β-glucosidase activity in the immobilized *A. faecalis* preparations. A micro-thermocouple (0.051 μm diameter) was used to measure directly temperature in gelling polyacrylamide slabs (7 mm $\times$ 2 mm $\times$ 0.2 mm). Activities were determined after disruption in a blender. Using acrylamide (750 mg) and BIS (30 mg) in 2.5 ml of water and β-dimethylaminopropionitrile with persulphate, reacting in a test tube, the temperature was shown to rise to cover 75 °C in the first 4 minutes, clearly detrimental to many enzyme systems. Using the slab system, with an aluminium block as heat-sink, was successful in keeping the temperature rise within reasonable limits, although the most dramatic effect probably arises from the presence of cells in the reaction mixture, which considerably slows polymerization. The greatest retention of activity was found under reaction conditions in which the greatest and most rapid rise and fall of temperature occurred (the maximum point not being above 40 °C) and was related by the authors to the different composition of these gels, which had the highest monomer concentrations. Provided temperature is prevented from reaching high levels, the diffusional properties of the gels are clearly of paramount operational significance. These results assist in the interpretation of the conclusions drawn by many workers regarding the unsuitability of polyacrylamide gels for particular cell types. Brodelius & Nilsson (1980), for example, found evidence of complete loss of viability (demonstrated by lack of cell growth or respiration or capacity to show plasmolysis) in plant cells immobilized in polyacrylamide.

Sarkar & Mayaudon (1983) compared entrapment in polyacrylamide with that in K-carrageenan and adsorption on DEAE (DE52) cellulose. Both viable and non-viable cells of *Corynebacterium dismutans* were used and alanine synthesis investigated. Thermostability was improved by polyacrylamide immobilization. The pH optimum for non-viable cells was changed from 9.0 (free state) to 7.5 when immobilized and the synthesis of alanine continued (on a daily batch basis) for considerably longer after immobilization. In a continuous column reactor polyacrylamide gave the best percentage synthesis of alanine, although inspection of their results shows this to be due primarily to high initial activities rather than more rapid inactivation with K-carrageenan or DEAE cellulose. Maximum activity in polyacrylamide gels was obtained using 0.33:0.020 of acrylamide: BIS (g in 3.0 ml). This was the lowest concentration of either reactant used and clearly reflects decreased toxity in the polymerization stage. Mechanical properties were not determined. The data obtained on apparent V_{max} and K_m values are difficult to interpret in the absence of information on particle size but the expected decreases in V_{max} and increases in K_m on immobilization were found.

Comparing polyacrylamide with K-carrageenan for the entrapment of microbial cells, Chibata (1979) reviews some industrial applications of both systems. The natural polymer is particularly recommended for the mildness of the entrapment conditions and the ease with which different shapes can be imparted to the biocatalyst particles. Although activity is higher, the mechanical properties of K-carrageenan are less satisfactory for unhardened beads. This point is returned to later in the present chapter. Bang *et al.* (1983) found that entrapment of *E. coli* in polyacrylamide was superior to polymethacrylamide or polyepoxide for the production of tryptophan from serine and indole. Retention of activity was 56% of free whole cells after immobilization, but the subsequent stability of the immobilized cells, in batch use, was considerably better. In these experiments the polyacrylamide gel was directly produced as beads by carrying out polymerization as suspended droplets in stirred phthalic acid dibutylester with a small amount of pluronic-L61 present and an atmosphere of nitrogen gas bubbled into the reaction flask. The temperature is not given, but the polymerization was continued for 20 minutes. For continuous production a CSTR was found to give promising results.

Comparing immobilization in alginate and polyacrylamide for the production of penicillin by *Penicillium chrysogenum* Suzuki & Karube (1979b) found less initial activity in the polyacrylamide gels but the poor mechanical properties of the alginate made it unsuitable in the presence of the necessary phosphate ions. Polymerization conditions in the polyacrylamide were adjusted by using high BIS concentrations. The half-life for penicillin production by the immobilized preparation was six times that of reused free mycelium. With bacitracin production from *Bacillus* sp. (KY4515) the concentration of BIS had no effect, but the overall production was no more than 25% of that of free bacteria. Regrowth was shown to occur by SEM observations. In the same paper the production of α-amylase by *Bacillus subtilis* immobilized in polyacrylamide is also reported. Regrowth was again observed and α-amylase increased with increasing numbers of reaction cycles. The rate of production was similar to that in free fermentation with a steady state being found after seven cycles. This example is particularly interesting for the large molecular weight of the product being studied.

Before leaving the topic of polyacrylamide entrapment it is worth noting an alternative approach to overcoming the cell damage caused by this technique. De Rosa *et al.* (1980) investigated the use of the thermophilic bacterium *Calderiella acidophila*, immobilized in polyacrylamide. The activity investigated was β-galactosidase. The preparation was stable for at least eight months of storage at 4 °C. The activity of the cells showed a nine-fold increase after immobilization, due apparently to permeabilization of the cellular membranes, but without complete cell lysis. Columns could be operated at 70 °C (a temperature above that allowing growth) and the β-galactosidase had a half-life of 30 days under continuous operation in the presence of the substrate, lactose. It is this development of organisms appropriate to immobilization, rather than the direct use of species or strains taken from conventional fermentation production systems, which offers great promise for the future development of immobilized systems.

Other synthetic polymers
The number of possible synthetic polymers which might be used for the entrapment of cells is very large (Table 3.2). Despite this, the majority of researchers have concentrated on the use of polyacrylamide. The use of other types of polymer is limited, probably due to the poor performance of many that have been tried. Klein, Hackel & Wagner (1979) compared a wide range of synthetic polymers for their ability to form ionic networks. Only copoly(styrene-maleic acid) was considered to show good gelling properties and aluminium sulphate was a satisfactory co-electrolyte. The natural-based carboxymethyl derivatives of cellulose and alginates were also satisfactory, although there appears to be no continued interest in the substituted celluloses as immobilization gels. A considerable number of calculations were carried out on results from the copoly(styrene-maleic acid) gels with various cell loadings, particle sizes and reaction temperatures using the degradation of phenol by *Candida tropicalis* as the test system. There appears to be no significant use of this synthetic polymer system by other researchers, probably because it cannot be used in the simple manner of alginates, which merely have to be weighed out from the bottle. The work of Klein & Eng (1979) with epoxy resins is discussed later in this chapter.

Fukui, Tanaka & Gellf (1978) prepared poly(ethylene glycol) methacrylate by refluxing PEG and methacrylate in toluene. Different chain lengths of PEG could be used and the methacrylate groups are attached at both ends. They also studied the analogue system known as ENT, made by reacting hydroxyethylacrylate with isophorone diisocyanate at 70 °C and a catalyst. After two hours poly(ethylene glycol)

Table 3.2. *Some synthetic polymeric gel systems used for entrapment*

Copoly(styrene-maleic acid)	Klein & Hackel, 1979
Poly(ethylene glycol) methacrylate	Fukui, Tanaka & Gellf, 1978
ENT	Fukui, Tanaka & Gellf, 1978
Methoxypoly(ethylene glycol) methacrylate	Fujimura & Kaetsu, 1981
Poly(ethylene glycol) dimethacrylate	Fujimura & Kaetsu, 1981
Polyisocyanates	Klein & Kluge, 1981
Polyurethane	Felix & Mosbach, 1982

was added and the reaction continued for a further five hours. The product in both cases is capable of photo-cross-linking. The type of PEG used is indicated by numbers after the respective letters, such as ENT-2000 when PEG 2000 has been used. Entrapment conditions can only be described as harsh, being carried out at 50 °C in the presence of 1% (v/v) of the initiator benzoin ethyl ether. After addition of the cell suspension (for example *Hansenula jadinii*), the mixture was sandwiched between sheets of clear polyester film and illuminated for three minutes with ultraviolet light. The 0.4 mm thick sheet of resin-immobilized cells was cut into small pieces for studies on its activity. They compared this method with polyacrylamide and glutaraldehyde cross-linked albumin entrapment. The unusual system for cytidine diphospho-choline synthesis was studied. This requires ATP from glycolysis, but is not shown by intact cells which convert cytidine monophosphate to uracil. Competent cells, produced by drying or detergent treatment, were successfully immobilized in ENT-2000 but not with polyacrylamide. The ratio of ENT-2000 to ENT-4000 influenced the rate of conversion of steroids by entrapped *Arthrobacter simplex*.

The production of ethanol by *Saccharomyces formosensis* was studied by Fujimura & Kaetsu (1981). They immobilized using methoxypolyethyleneglycol methacrylate (MPEGMA), or polyethyleneglycol dimethacrylate. Polymerization was achieved with gamma-radiation from a Cobalt-60 source (1×10^6 rad/h) for one hour. The polymer was supported on rolled gauze. As expected, many but not all cells were damaged by this treatment. Regrowth was possible but much improved by lower irradiation (0.5×10^4 rad/h) and at low temperatures (e.g. -24 °C and 0 °C). The monomer of MPEGMA was somewhat less toxic than that of 2-hydroxyethyl methacrylate (HEMA) and gave comparable viability to untreated controls. Kumakura & Kaetsu (1983) used *Streptomyces phaeochromogenes* immobilized in a 10% HEMA matrix using the same Cobalt-60 source but an irradiation temperature of -78 °C, carried out in a glass tube. The polymerized mass was cut into small flakes. Before immobilization the cells were surface-coated (30 seconds) with a wide range of hydrophobic monomers: hydroxybutyl methacrylate (HBMA), hydroxyhexyl methacrylate (HHMA) and tetraethyleneglycol acrylate (A-4G). The organic liquids *n*-hexane, *n*-octane and carbon tetrachloride were also tested. These gave significant difference in the retention of glucose isomerase over twelve batch reactions. The pre-treatments with HHMA, HBMA and A-4G were best, this being attributed to prevention of cell leakage from the matrices. Using 50% HEMA alone, there was much less activity detected, probably partly due to the small pore size. The coating of monomers appears to cause retention of the cells even in the 10% HEMA gels which had pores of around 20 μm diameter. Clearly, the subtle modification of techniques can overcome apparent problems, such as cell leakage from large pore gels.

Klein & Kluge (1981) selected the hydrophilic polyisocyanates as prepolymers for the immobilization of *E. coli*. These can be used both as gels and as foams. Details are given for the production of these and spherical foam beads (made in stirred liquid paraffin), although different types of polyisocyanates may be required. The gas bubbles are carbon dioxide and the gel is formed by having a large amount of water present to dissolve the gas. For the foams the volume increases rapidly in the first few minutes to some five times the original. Temperature rises are small and the pH value does not change. Density and porosity are controlled by the amount of water present. The pressure during polymerization can also influence the polyurethane foam density. Rigid, low density foams appear to have the best properties. Felix & Mosbach (1982) immobilized cells of the higher plant *Catharanthus roseus* in polyurethane foam and compared it with agarose, alginate and a glutardialdehyde cross-linked gelatin foam. Only the cells in agarose and the polyurethane (hypol 3000) showed both isocitrate dehydrogenase and cathenamine reductase activities. Treatment with dimethyl sulphoxide is necessary to allow detection of resonable activities of the isocitrate dehydrogenase. The ease and rapidity of using the polyurethane foam was emphasized. Egerer *et al.* (1982) showed good retention of the ability to produce $NADH.H^+$ when cells of *Alcaligenes eutrophus* were immobilized in a polyurethane gel (PU-6). Immobilization was also carried out in ENT-2000 but this did not appear as satisfactory as the PU-6 system.

Mohamed & Salleh (1982) studied the physical properties of a gel matrix formed from polyethylene glycol alginate and polyethyleneimine, a mixture which gels very quickly. The properties of the gels can be controlled by the suitable choice of conditions and the authors suggest a distinct advantage over protein-containing matrices, because of the inability of the proteases to disrupt the gel system. Other developments of this and the almost infinite range of possible immobilization matrices will certainly occur.

Celluloses
Although cellulose itself is insoluble in water it can be dissolved in certain organic liquids. Linko, Phjola & Linko (1977) dissolved cellulose in *N*-ethyl pyridinium chloride and dimethyl formamide for the entrapment of *Actinoplanes missouriensis*. Beads are formed by dropping into water. The target enzyme, glucose isomerase, showed between 40% and 60% of its original activity after immobilization. Glutaraldehyde was used as a cross-linking agent to prevent cell leakage. The half-life of the preparation was 45 days and one may reasonably conclude that the cells were non-viable after the entrapment. Although a few other species have

been immobilized by this technique, it is not generally considered to have any advantage over the various available alternatives. A few researchers have tried cellulose acetates for entrapment but there generally appear to be problems of leakage and diffusional restriction, with no particularly beneficial features. Sakimae & Onishi (1981) have patented a method involving cells protected from freezing damage by various cryoprotectants. Beads of frozen cells were formed by dropwise addition to n-hexane at temperatures down to $-70\,^{\circ}\text{C}$. The beads were separated and added at low temperatures to solutions of a variety of entrapment polymers, including cellulose acetate. Ice and solvents were then removed by freeze-drying. It is unlikely that such a cumbersome method will be widely adopted unless real advantages can be demonstrated.

Dinelli (1972) gave a thorough discussion of the entrapment of both enzymes and cells in solid fibres made from a variety of polymers, particularly those based upon celluloses. This has the advantage of using standard equipment for the wet spinning of fibres and can clearly be scaled up easily to any desired quantity. Threads could be woven into fabric . When *E. coli* were trapped in cellulose acetate they showed approximately 80% (compared to free cells) of penicillin acylase activity, when used at the rather low loading of 15 mg wet cells/g polymer. More heavily loaded fibres showed very reduced activities. *Saccharomyces lactis* at 75 mg dry wt cells/g polymer (cellulose triacetate) showed 10% of the β-galactosidase activity of free cells. The standard procedure involves cellulose triacetate dissolved in methylene chloride but a suitable combination of the developments made by Sakimae & Onishi (1981), mentioned previously, to reduce solvent damage, could lead to useful developments. The main block to these developments is the fact that equipment and expertise to spin fibres are not readily available in laboratories carrying out immobilization studies, plus the preference for spherical beads shown by many researchers. Bioactive membranes can also be produced for use in standard ultra-filtration reactors, as noted in Bucke (1983).

Joshi & Yamazaki (1986) used cellulose acetate in acetone, mixed with *E. coli* cells, to impregnate cotton cloth. The immobilized cell cloth was then treated with polyethyleneimine and glutaraldehyde to stabilize the target aspartase activity. The cell loading and porosity of the cellulose acetate aggregated onto the cloth could be adjusted for optimal aspartase activity. A complication due to the liberation of CO_2, probably due to the action of malic acid decarboxylase, could be overcome by the use of sodium dithionite. This probably prevented the regeneration of the oxidized form of the coenzyme NAD, needed for continued decarboxylation. Comparative experiments showed that an STR gave 2.5 times faster production of aspartic acid than a packed column but it is not to be expected that 0.5 cm squares of the cell cloth would be an optimal configuration for either type of reactor.

Two-step methods

From what has been said so far in the present chapter about the properties of immobilization matrices, it is obvious that a combination of alternative methods may well provide a good means of overcoming evident shortcomings. This approach has been tried by several researchers. Alginate gels are clearly prepared under mild conditions and would provide a possible protection to the cells while more stable matrices are being created under physiologically harsher conditions. Klein & Eng (1979) combined alginate gells with an epoxy-resin matrix, using *E. coli* cells and their penicillin acylase activity as a marker of success. The wet cells were mixed with the epoxy resin reagent (in this case 'Epikote ^{R}DX 255') and curing agent ('Casamide^R') in an aqueous medium. To this was added sodium alginate and the suspension was added to calcium chloride, dropwise, to give conventional calcium alginate beads. After allowing time for the polycondensation of the resin the beads were air-dried and the alginate dissolved out in a phosphate buffer (Figure 3.6). During this process the beads swell from their dried size and an open porous structure is formed. These have greatly improved properties over particles of the same diameter (3 mm) made by single-step production with epoxy-resin, followed by grinding. The epoxy-resin method itself is mild, as shown by the high activity obtained when single-step particles of 100 μm are assayed. These small particles actually have higher relative activity than the larger ones from the two-step process, but are less suitable for packed or fluidized-bed use. The two-step beads also showed more elastic behaviour and are not so easily damaged by mechanical handling. In a subsequent paper Klein & Kressdorf (1982) described an improved method. Although the original method did allow the retention of a certain proportion of viable *E. coli* cells, it did not allow yeast cells to remain viable. They carefully selected types of resin and curing agent which had low contents of low molecular weight components (using for these experiments 'Eurepox^R 730' and 'Versamid^R 115'. These were allowed to react together for 15 minutes before adding the yeast cells (*Saccharomyces cerevisiae*) already suspended in sodium alginate solution. Beads were then formed as described before. An initial alcohol production activity (of 21% of the original free cells) was obtained only when the pre-condensation stage was used. The beads could be reactivated under growth conditions and had good mechanical properties.

Alternative approaches involving the stabilization of alginate beads have been studied by various researchers. Birnbaum *et al.* (1981) compared three methods of stabilization. Firstly, conventional alginate

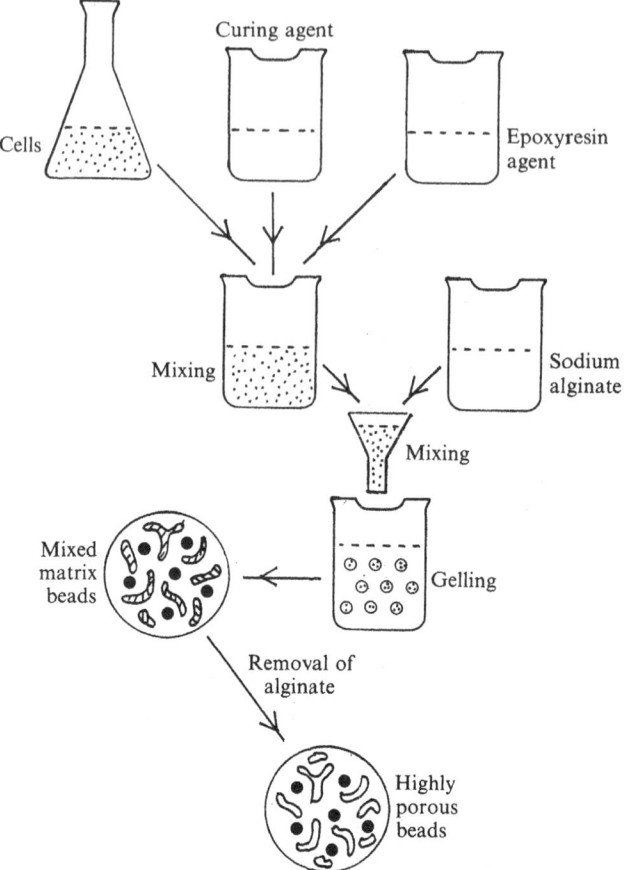

Figure 3.6. Diagrammatic representation of the principle of two-step gel entrapment using epoxy-resins.

beads were sequentially treated with polyethyleneimine-HCl and glutaraldehyde, the former being allowed to infiltrate the beads for 24 h while the latter was applied as a pH 7, 1% (v/v) solution for one minute. The beads were then washed in water. In the second method the alginate was first activated for one hour with a mixture of N-hydroxysuccinimide and 1-ethyl-3-(3-dimethyl-amino-propyl)-carbodiimide. After addition of the cells, beads were formed by conventional addition to calcium chloride solution and allowed to cure for one hour. Beads were then stabilized by a 24 h treatment in a defined suspension medium containing polyethyleneimine, and washed in water. For the third method the sodium alginate was pre-reacted with sodium meta-periodate for one hour. The cells, in an equal quantity of unreacted sodium alginate, were then mixed in and beads formed. After curing for one hour the beads were treated with polyethyleneimine-HCl and washed in water. The exact mechanism of all three methods is not fully known, but they all probably involve some ionic or covalent binding of the cells themselves. The authors consider it might even be appropriate to describe the second and third methods as involving an element of direct coupling of the cells to the activated alginate. The properties of the glutaraldehyde-stabilized beads could be varied according to the conditions used for the polyethyleneimine-HCl infiltration. There was possible evidence of some toxicity by the glutaraldehyde to the yeast (*S. cerevisiae*) immobilized, as shown by somewhat reduced ethanol production, but this method still retained considerable activity and was much cheaper than the two alternatives tested, which both yield active preparations. This paper contains much interesting discussion and further developments with these techniques seem certain.

SivaRaman *et al.* (1982) used a mixed alginate–gelatin system for yeast (*S. cerevisiae* and *S. uvarum*) immobilization. Cells were added to a mixed (20% (w/v) gelatine and 2% (w/v) sodium alginate) polymer solution and beads formed by dropping into calcium chloride solution. The alginate was then leached out in phosphate buffer and the porous gelatin beads stabilized by cross-linking with glutaraldehyde. With careful control of the glutaraldehyde concentration enhanced fermentation rates can be obtained (compared to free cells), but use of greater than 0.015 M glutaraldehyde caused reduced rates. Good ethanol productivities were obtained in a continuous packed-bed reactor and some budding observed. The use of this method of immobilization formed part of a process for the simultaneous saccharification and fermentation of cellulose described by Deshpande, SivaRaman & Rao (1983).

Chibata (1979) reported on the stabilization of K-carrageenan beads by treatment with hexamethylenediamine (HMDA) and glutaraldehyde. The greatest extension of the half-life of *E. coli* aspartase activity (to 680 days) occurred with these two reagents, each used at 85 mM, although there was negligible difference in the initial units of aspartase activity if the glutaraldehyde was used at lower concentrations. The extension of half-life was much better than when using glutaraldehyde alone or persimmon tannin as hardening agents or using polyacrylamide immobilization. Productivity was 15 times higher with the HMDA plus glutaraldehyde-hardened K-carrageenan compared to polyacrylamide.

Immiscible liquid systems

The use of a solid support or an entrapment matrix for cell immobilization leads inevitably to certain problems. The support material itself may carry charges which, although desirable for the process of immobilization, may adversely affect the expressd enzyme activity of the cells. This may occur, for example, by an effect on the local pH value around the biocatalyst. Another major problem is that of restricted diffusion. This is manifested in two major ways. Firstly, the supply of substrate to the biocatalyst will often not be sufficient to maintain optimal rates of bioconversion. Secondly, the products of the reaction may also have difficulty in diffusing away from the reaction site and this will lead to product inhibition of the biocatalytic activity. The former problem has received most attention because it is easier to carry out experiments and calculate projected rates of biocatalysis. One approach to this problem is to dispense with a solid support and instead entrap the cells within one liquid phase which can be reduced to very small droplets by emulsification within another liquid phase. Allowing the emulsion to break will enable the cells to be recovered readily from the products. Clearly, two features are of paramount importance in such systems. Firstly, the target biocatalytic activity must be stable in the two-liquid system chosen and secondly, the partitioning of any necessary nutrients, the reactants and the products, must be appropriate to the particular situation under consideration. Fortunately, there is an enormous literature available on the partitioning of all manner of substances, from low molecular weight substances, through macromolecules and organelles to whole cells. As early as 1971, Albertsson was issuing the second edition of his book dealing with the partitioning of macromolecules and cellular subfractions. Since then the use of two-phase systems has become a standard part of the recovery of protoplasts and even intact whole cells.

Liquids can be placed, with very little disagreement between scientists, into a rank series of increasing hydrophobicity. Those lying close to water in the series, such as ethanol and acetone, can be mixed in any proportions with water to give a single-phase solution. Moving in the direction of increasing hydrophobicity the amount of water which can be present (generally expressed on a volume : volume basis) decreases so that two liquid phases occur more frequently, until the organic liquid is essentially totally immiscible with water. It should be pointed out that water is not the opposite end of the series because solids, particularly inorganic salts, dissolved in water produce a solution with greater affinity for water than pure water itself.

The main problem with using organic liquids as the second component in a two-liquid-phase system is that, because of the hydrophobic nature of lipid components in the cell membranes, they are almost certain to alter the selective permeability of such membranes. Indeed, in cell immobilization technology they are often specifically used to permeabilize the cells and allow easier access of substrates to sites of biocatalytic activity. This may be perfectly satisfactory for many applications of immobilized cells, but will inevitably lead to loss of cell viability. Fortunately, an alternative and very mild method of producing a two-liquid-phase system exists. This is to use solutions of water-soluble polymers. Because these polymers can be selected to have different degrees of inherent hydrophobicity, while still being water-soluble at appropriate concentrations, it is possible to choose a combination of two, separately prepared, solutions of such polymers which are not miscible. This is possible even though both phases contain over 80% water and the surface tension between the two aqueous phases is extremely low. Mattiasson & Hahn-Hagerdal (1983) present a review of the use of such aqueous two-phase systems and the present brief discussion is given merely to highlight certain interesting ideas arising from the use of such systems.

One problem of using cells immobilized in gels or on solid supports is that such biocatalysts are clearly unsuitable for use with particulate, water-insoluble substrates such as cellulose. Although the cellulase complex enzymes are generally secreted by producing micro-organisms, the end products of cellulose hydrolysis generally exert a strong inhibitory influence on the rate of reaction. Many researchers have therefore sought to remove such end products as quickly as they are formed by bioconversion to substances which do not directly inhibit cellulolysis. Hahn-Hagerdal *et al.* (1981) used a two-phase system of Dextran T-40 and Carbowax PEG 8000 to study the enzymic hydrolysis of cellulose, coupled to the use of the liberated glucose by *Saccharomyces cerevisiae* and the production of ethanol. They found considerable advantages in using the two-phase system compared with a single-phase system. The authors also point out the simplicity of adding more of the least stable biocatalyst, while the complete system is still operating. This is something which is not readily possible with gel beads or other solid immobilization supports.

One of the major problems with the production of solvent ethanol from low-cost substrates is the recovering of the ethanol from the system. The selection of a suitable two-liquid-phase system, with appropriate ethanol partition characteristics could allow a continuous process to operate with the ethanol being continuously distilled from the non-cell phase, which is then recycled back to the reactor. Kuhn (1980) used a Dextran 500 phase and a PEG 6000 phase in this way, with glucose being fed to *Zymomonas mobilis* but on a fed-batch recycle rather than fully continuous operation. After a number of recycles the ethanol production rate fell, for reasons which were not fully discovered. An obvious development in solvent ethanol production, which has been discussed by many researchers, is to use thermophilic organisms, with the advantages of lower risks of contamination and the need for only a small additional heat energy input to recover the ethanol. Ljungdahl (1983) is reported to have used *Thermoanaerobacter ethanolicus* in a two-phase system at 70 °C. The partition coefficient of the ethanol in the system used (polyethylene glycol- and dextran-containing phases) was found to differ significantly between room temperature and 70 °C, with the latter coefficient being much smaller. It is important to keep the ratio as high as possible between the non-cell phase and the cell-containing phase to reduce product inhibition to a minimum. In many cases, of course, the substrate and products will be equally soluble in both of the aqueous phases and this stricture will not apply.

An interesting aspect of the removal of inhibitory substances from the immediate vicinity of producing cells is the possibility of obtaining greater yields of toxic metabolites. If the volumes of the two aqueous phases are deliberately set to be widely different, the amount of toxic substance present in the non-cell-containing phase can considerably exceed the amount possible with a conventional one-phase system. Puziss & Heden (1965), for example, obtained increased yields of *Clostridium tetani* toxin in this way, using dextran- and polyethylene glycol-based phases in a 1 : 15 volume ratio. These systems are useful even when the partition coefficient of the product is close to unity in the two phases. The possibility of including a specific binding group (ligand) in only one of the two phases could lead to positive accumulation of the target product into that phase even where the partition coefficient itself was not particularly favourable.

A major difficulty with the use of water-soluble polymers for two-phase systems is that many of the systems have been originally designed for biochemical separation purposes, where the cost factor was not of importance. In an industrial setting it may be imperative to find cheap, readily available polymer systems. Specific dextran fractions are expensive, but the related crude, polymeric complex starch, is cheap and can also serve as a substrate itself. Studies have been made of such systems using a heat-stable amylase to hydrolyse the starch, at various temperatures up to 80 °C. The system has been studied as a single-stage saccharification process and with a coupled ethanolic fermentation.

The use of two-phase systems in which only one of the phases is water-based has also been investigated. This has been used predominantly for reactions involving a substrate or product that is not readily soluble in water. A typical example of this is the bioconversion of cheap precursor steroids into derivatives which are more pharmacologically useful. Because of the general toxicity of the organic solvent phase to cells, much of this research has involved separated enzyme rather than cell-based conversions. Antonioni, Carrera & Cremonesi (1981) discussed such enzyme-catalyzed reactions. As Lilly (1983) pointed out, two-phase systems can be used even where the biocatalyst is not itself particularly stable or active in the organic phase. In some cases it may be that a certain amount of toxicity by the organic solvent is offset by a higher solubility of the substrate in that solvent.

Thus, in the enzymic conversion of cortisone to preg-4-en-17,20,21-triol-3,11-dione the preferred organic phase is ethyl or butyl acetate-based, despite around 70% and 50% enzyme inhibition respectively, because of the much greater solubility of cortisone in them when compared with less toxic solvents, such as chlorobenzene. Another benefit from using an organic solvent is that an enzyme which is predominantly hydrolytic in an aqueous medium can be encouraged to bring about the opposite condensation reaction. Strobel *et al.* (1983), for example, used enzyme self-immobilized in dried mycelium of *Rhizopus arrhizus* to synthesize esters of pentanoic, butyric and acetic acids. In this case various reacting alcohols were themselves used as the solvents. The same fungus was used by Bell *et al.* (1978) to synthesize mono- and diglycerides, in an acetone solvent system. Such potentially useful developments are reviewed by Fukui & Tanaka (1982). The specificity of biocatalytic reactions may be considerably changed by carrying out the reactions in organic solvent systems.

An emulsion of *Nocardia corallina* cells in liquid paraffin, with an added antifoam agent, was used by Miyawaki *et al.* (1986) to investigate propylene oxide formation. A bubble-lift reactor was used. Both the biocatalytic activity and the stability were improved by this liquid immobilization system, with the cells preferring the hydrophobic phase. As might be expected, the influence of pH value was small but there was evidence of some diffusion limitation with higher gas flow rates giving greater apparent activity.

Microencapsulation

The technique of microencapsulation is well established and has a wide range of applications. A selectively permeable or impermeable membrane is created around droplets or particles to prevent the free diffusion of pre-determined substances either in or out of the inner compartment. In some cases the membranes may be impermeable under one set of conditions but become permeable or disrupt under another. Typical examples of this are the temperature-dependent encapsulation of enzymes for use in biological washing powders or pigments and essential oils in heat or touch sensitive microcapsules. In theory it can be seen that the use of small membrane-bounded spheres, freely dispersed in the liquid phase of a reactor, will provide a high surface area for exchange between the cellular microenvironment and the bulk liquid phase and might be expected to give easier separation of the cellular phase from the bulk liquid.

Unfortunately, several practical problems have limited the use of encapsulation. Firstly, the solvents in which the membrane raw materials are dissolved and even the membrane precursors themselves are often highly toxic. Where a viable cell preparation is needed to carry out complex bioconversions or total syntheses there will be an inevitable loss of catalytic activity. Secondly, where the problem of toxicity has been overcome there may be problems due to cell growth, division or gas production causing mechanical rupture of the encapsulation membrane. Where non-viable cells are being used these problems will not exist, but a third one applies to both types of preparation. This is the additional diffusion limitation which results from the presence of the membrane. While it is true that this will not be as serious as with some other types of immobilization, it will be greater than when the cells are immobilized onto similar-sized support particles without an exterior membrane. In the case of simple encapsulation of just cells within the membrane, a fourth problem arises from the fact that the preparation then has a density little different from the bulk liquid and therefore retention of active biomass in a continuous reactor may become a problem. This would limit the range of reactor designs which could be used and the substrate flow rates. This could be overcome by co-encapsulating a denser material with the cells, but workers have preferred to immobilize directly onto such particles.

Another interesting aspect of this is the possibility of using an absorbant for the products of the reaction within the microcapsule itself. Miyawaki, Nakamura & Yano (1978), for example, have modelled the mass transfer and kinetics for the degradation of urea within microcapsule, containing an absorbant (ion exchange resin) for the ammonia produced during the reaction. In work with *Micrococcus denitrificans* Mohan & Li (1975) prepared microcapsules of 20–40 μm diameter each containing 500–600 cells. Despite the emulsion-like properties of the preparation some diffusion limitation was still found, but a reasonable retention of biocatalytic activity was obtained and no cell lysis was seen after five days,

Obuchi & Maeda (1985) developed a microcomputer system to control the flow reversal in a horizontal column reactor containing microcapsules. Pressure drop was sensed by a digital manometer and used to activate flow reversal. The strategy and programmes of the type used in this application could easily be adapted for other situations and so allow the more efficient use of compressible support systems. Hartmeier & Heinrichs (1986) immobilized *Gluconobacter oxydans* in alginate, with added purified fungal catalase. The mixture was sprayed into an emulsion of acrylate-methacrylate polymer. This yielded membrane-coated beads of 1 mm diameter which retained the catalase activity and aeration could be improved by using hydrogen peroxide as the oxygen donor. This technique is a useful approach to the problem of diffusion limitation in gel beds.

Immobilization by permeable membranes

The use of selectively permeable membranes to retain macromolecules, but allow the free diffusion of low molecular weight solutes, is one of the oldest biochemical techniques. At first, the membranes were taken from natural animal sources but these are difficult to store and handle. For a long period of time the best replacement for these was membranes made of regenerated cellulose and these are still extensively used because of their cheapness, ease of preparation and efficient operational characteristics. Major advances in membrane technology have occurred more recently, because of demands from a variety of users. In medical applications the development of dialysis machines has now reached a high degree of sophistication. In biochemical research attention has focused particularly on ultrafiltration, which involves the use of membranes of specified pore size ranges to separate and frequently concentrate macromolecules (particularly proteins) from low molecular weight contaminants. The major advance here has been to operate pressure- (or less commonly vacuum-) driven flow through the membranes, thereby speeding up the concentration stage considerably. An alternative approach is to increase greatly the surface area of the membrane and reduce the diffusion path for those molecules which can pass through the membrane pores. This has led to the development of hollow-fibre devices, with mass flow of liquid, in separate streams, both inside and outside of the hollow fibres. The individual hollow fibres are mounted in such a way as to allow a single large inlet to feed a multiple bundle of 100 or more of selectively permeable hollow fibres, with a corresponding system to recombine the outlet streams. Although some of these are still based on cellulosic membranes, a number of entirely synthetic membranes are also now used. Clearly, such systems are not cheap to manufacture and users therefore look for an extended lifetime for these sophisticated ultrafiltration devices. The use of such systems to retain cells is an obvious extension of their use to retain and concentrate macromolecules and to entrap enzymes used for biocatalytic reactions involving low molecular weight substrates and/or products. Some membrane separation systems are shown in Figure 3.7.

Although it is obvious that retention by artificial selectively permeable membranes is the mildest possible way of immobilizing cells, there are a number of operational problems to be overcome. Firstly, the pore size must be appropriate to the application. It is not possible to produce pores which are all exactly the same diameter and path length. Attention therefore focuses upon the approximate upper limit (the cut-off point or exclusion limit) size of the pores. For ultrafiltration membranes these are generally quoted in molecular weight units (or daltons) rather than actual pore diameters and, because of the range of macromolecular shapes

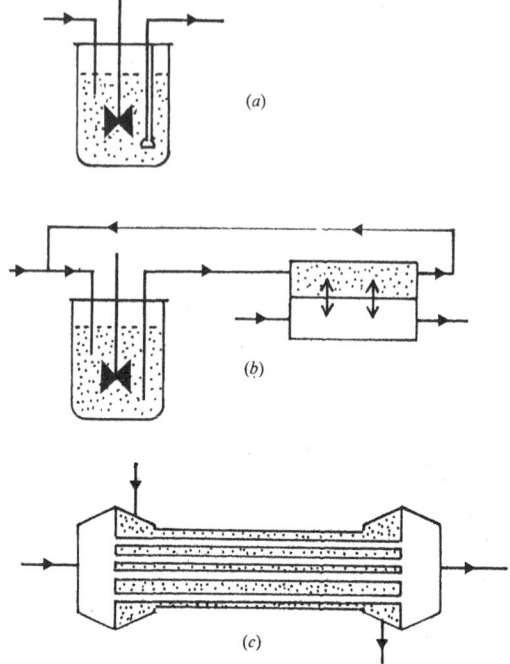

Figure 3.7. Diagrammatic representation of three types of membrane reactor system for cell immobilization. (*a*) Membrane separating the outlet stream from the reactor contents; (*b*) external recycle loop with membrane separation unit; (*c*) hollow-fibre unit used as a single unit membrane reactor.

which can correspond to a given protein molecular weight, plus the influence of variable pore diameter, a range is given rather than a single figure. At this point it should be emphasized that most membrane systems have been developed for applications involving the separation of macromolecules from low molecular weight solutes rather than the separation of cells from macromolecules. This is, perhaps, less of a problem than it might be, because in many immobilized cell applications it is important to keep secreted macromolecules (such as extracellular enzymes) in the same compartment as the biomass and only allow low molecular weight nutrients and products to pass into the other stream. There is, nevertheless, considerable scope for the development of membranes with larger pore sizes, but in the configuration of hollow-fibre devices or other large-scale systems. One interesting approach is to modify the growth form of the organism to allow the use of membranes with larger pore sizes. Tso & Fung (1981) used sublethal concentrations of Crown ether to induce a filamentous morphology in the rod-shaped bacteria *Escherichia coli, Salmonella typhimurium* and a species of

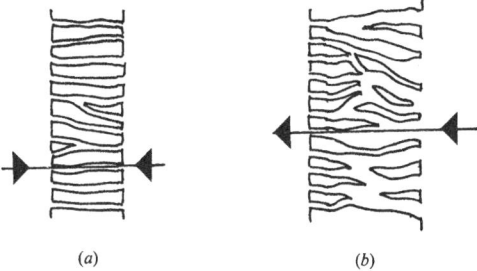

(a)   (b)

Figure 3.8. Diagrammatic representation of the two major types of membrane used in membrane reactors. (*a*) Isotropic membrane with both faces having pores of the same size; (*b*) anisotropic membrane with the outer face having much larger pores leading to the small pores that effectively control the size of the particles passing through the complete membrane.

Bacillus. No change in morphology was found with the species of *Streptococcus, Staphylococcus* and *Corynebacterium* tested. Where effective, the cell lengthening induced by 18-Crown-6 was approximately proportional to its concentration. They do not, however, go on to describe a suitable membrane retention system for the filamentous forms and, in view of the mammalian toxicity of Crown ether, it might be expected that there would be some resistance to its adoption.

Major problems with any selectively permeable membrane systems are how to keep the pores from becoming clogged and how to sterilize the system, which is essential for many applications. The original membranes were isotropic, having both faces of the membrane the same and pores of, approximately, equal diameter running from one side to the other. Because of their improved operational characteristics many applications now use anisotropic membranes or hollow fibres. These consist of a thick (say 50–75 μm for hollow fibres), highly porous membrane to which is attached a thin, defined pore size, ultrafiltration membrane, usually lining the internal lumen of the fibre (Figure 3.9). For a detailed description of the available types of membranes and ultrafiltration systems the reader is referred to the relevant trade literature from the manufacturers. There is also a voluminous literature on the use of such systems in applications which do not involve cell immobilization. In general, membrane systems based on cellulosic materials are difficult to clean, other than by conventional back-flushing against the normal direction of flow. Many cells adhere to the membrane itself and are difficult, if not impossible, to remove. Synthetic polymer membranes are manufactured which have a much greater stability to a wide range of conditions. Van Altena (1975), for example, gives some results for cleaning non-cellulosic hollow fibres by a variety of techniques, including a very effective regime of 1% (w/v) sodium hydroxide at 50–55 °C. The experiments did not, however, include fouling by cell growth.

One of the original uses of hollow-fibre units was in the culture of mammalian cells. Examination of natural tissues revealed that healthy cells were always within about $50\,\mu$m of a blood capillary and that, even in neoplastic tissue, cell death occurred when the distance was greater than $150\,\mu$m. Based on this observation a culture system was devised which allowed the passage of culture medium along the lumen of hollow fibres. The cells were cultured in the extracapillary space, the whole unit being maintained under suitable environmental conditions. The use of such systems for the culture of mammalian cells was reviewed by Hopkinson (1983), of the Amicon Corporation, USA, who supply this type of unit. Although numerous experiments have been carried out to investigate optimal configuration, construction materials and operational conditions, most units now operate with anisotropic fibres of polysulphone or acrylic copolymers and supply oxygen via the serum-containing culture medium in the lumen compartment. One of the great advantages of using hollow-fibre culture, both for animal and other cell types, is the very high density of cells which can be achieved. There is also a considerable saving, in the case of animal cells, in the amount of high-cost culture medium needed, because the macromolecular components are restricted to the intracapillary volume. Further advantage comes when high molecular weight products from the cells are being recovered because these, by the same mechanisms, are retained in the extracapillary fluid and so are not contaminated with serum proteins. A wide range of animal cell types, including normal and anchorage-dependent cells, as well as cancerous cells, have been grown in hollow fibre systems. Because the cells do penetrate the larger outer pores of the fibres it is, of course, impossible to recover completely all the cells present. The systems can, however, be operated for extended periods, with batch or continuous removal of excess cells from the extracapillary vessel. A certain amount of functional differentiation can occur in hollow-fibre systems, with continued liberation of target products even in the absence of continued cell division. As Hopkinson (1983) points out, there is a great potential for the use of such systems which is not yet being fully exploited. This is partly due to the high cost and extensive expertise needed to operate such systems and partly due to the fact that much of the work which has been done is not available in published form, for commercial reasons.

It is not essential to use expensive hollow-fibre devices to carry out useful work on the culture of membrane-retained animal cells. Adamson *et al.* (1983) used ordinary (9 mm diameter) dialysis tubing which had been acetylated and pre-treated before autoclaving. Mouse hybridoma cells capable of producing monoclonal human transferrin antibodies were cultured inside the dialysis tubing in small-scale (1 ml) batches. A dialysable, low molecular weight, component of foetal calf serum was shown to give a ten-fold increase in cell density and antibody titre. The internal to external volume ratio of 4:31 appeared sufficient to prevent the accumulation of harmful levels of lactate and the serious depletion of glucose which occurred in ordinary batch cultures of the hybridoma cells. The use of a commercial-scale, hollow-fibre system, with a similar operational regime, would be an obvious extension of these research findings.

Turning now to microbial systems involving the membrane retention of cells, an obvious candidate for study is the production of solvent ethanol. For economic reasons it is necessary to have a high cell density giving a high volumetric rate of production and yet to minimize the toxic effects of the product. Membrane retention is only one possible mechanism for obtaining a biomass recycle system. Cheryan & Mehaia (1983) used a cross-flow membrane system coupled to a fermenter in a semi-closed loop configuration. Using *Kluyveromyces fragilis*, grown on a synthetic lactose-containing medium an 80-fold increase in productivity was claimed, when compared to a batch fermentation. A standard fermenter was coupled to a PM-50 hollow fibre cartridge device (Trade Mark of Romicon Inc., USA) with a stated 50 000 MW cut-off. The highest ethanol productivity of 240 g/l/h was obtained with a cell concentration of 90 g dry wt/l, a dilution rate of 6/h and a lactose feed concentration of 150 g/l. Favourable comparison is drawn with other systems using membrane retention methods, such as that of Lee *et al*. (1980). The latter researchers used two strains of *Zymomonas mobilis* and a glucose-containing feedstock. They also compared cellulose acetate membranes (pore size 0.2 μm) in a tangential-flow microfiltration unit with polyamide membranes (pore size 0.2 μm) in a cross-flow capillary microfiltration unit. They experienced significant reductions in the permeability of the membranes, particularly with one of the two strains of *Z. mobilis*. This was not fully explained but was suggested to be due to overgrowth of the membranes and possible effects of the ethanol on the cellulose acetate membranes. A configuration using parallel polyamide membrane units, which could be alternately switched in and out for cleaning with 1% sodium hydroxide was adopted. High productivity levels could be maintained by switching units after between 12 and 15 hours of operation. These researchers also noted the need to develop an appropriate predictive model for membrane operation where gas bubbles, arising from metabolic activity, cause additional mass transfer resistance through the membrane. In their work Chen & Zall (1982) deliberately chose cellulose acetate as the cell support material for a continuous attached film expanded bed reactor using *Saccharomyces fragi* to ferment ultrafiltered whey to ethanol. This was on the basis of previous observations of the attachment of this organism to cellulose acetate ultrafiltration membranes. The support was not ideal because gas production caused particles to float and be carried over from the reactor. In an alternative fermenter configuration reported in the same paper they also included a simple membrane unit. They did not, however, comment upon fouling of the membrane unit but did suggest a combination of their two systems.

Some of the problems of hollow-fibre units were encountered by Roy, Blanch & Wilke (1982). They used an Amicon Vitafiber (Trade Mark Amicon Inc.) containing polypropylene hollow fibres (Celgard, Trade Mark of Celanese Fiber Company, USA). The hollow fibre unit was sterilized with ethylene oxide. The target product was lactic acid with *Lactobacillus delbrueckii* in the outer compartment of the unit. A recycle system was operated from a reservoir through the lumen of the hollow fibres. High cell densities were obtained, sufficient for the mass transfer resistance in the biomass compartment to be greater than that through the membrane. After approximately 100 hours the cell mass had distorted and penetrated the fibres in their 300 fibre unit. The biomass developed was so concentrated (480 g dry wt/l) it was suggested that part of this must consist of dead cells or cells with a much below normal water content. It is clear that in these experiments the biomass increased to such an extent as to impair performance, although high volumetric productivities were obtained (100 g lactic acid/l/h). It would appear that the hollow fibre system which works so well with slower growing animal cells is less suitable for strongly growing bacteria. Roy *et al.* (1983) carried out further experiments with lactic acid production by *L. delbrueckii*. For these they used a conventional CSTR coupled with an external membrane filtration unit. This unit was a Pellicon Cassette System (Trade Mark of Millipore Corp.) fitted with polysulphone membranes having a cut-off point of 100 000 MW. In these experiments the filter unit was treated with 3.7% (w/w) formaldehyde solution before use, to reduce microbial contamination risks. An alternative method, using 70% (w/w) ethanol solution had previously been found to be less satisfactory. The volumetric productivity was 76 g lactic acid/l/h with a biomass concentration of 54 g dry wt/l. The membranes performed well over an extended period, with flow direction reversed every few hours and cleaning between experiments with 0.2 M sodium hydroxide. Their comparison of the results with other studies on lactic acid production clearly showed the superiority of both their present and previous systems in terms of volumetric productivity. Their own preliminary economic analysis, not fully detailed in the publication, indicated a price advantage over batch fermentations for lactic acid. Mattiasson *et al.* (1981) compared a hollow-fibre reactor with an alginate entrapped system using *Pseudomonas denitrificans* for the continuous denitrification of water. An upflow packed-bed column reactor was used for the alginate beads. The hollow-fibre unit was a Lundia fibre (supplied by Gambro AB) with the free bacteria in the outer compartment. For comparable amounts of cells the hollow-fibre system gave a greater reduction in nitrate (fed at 100 mg/l) than the alginate bead system, indicating less diffusion restriction in the former system. Cell leakage was consistently lower from the hollow-fibre reactor. Gas production was a problem with both reactors but was overcome easily for the hollow-fibre device by circulating the bacterial suspension from the outer compartment to an external unit, containing a hydrophobic membrane that allowed the gas but not the liquid phase to escape.

The transformation of benzaldehyde to benzyl alcohol was studied by Wisniewski, Winnicki & Majewska (1983) using a polysulphone membrane formed onto a sintered polyvinylchloride support. This formed the entrance to an inverted cone inserted into a CSTR. The effective surface area of the membrane was just over 10 cm^2 in a reactor volume of 180 cm^3. The organism used was *Rhodotorula mucilaginosa*. Nitrogen gas was used to pressurize the CSTR and provide the motive force for membrane filtration. The organism is unable to use benzaldehyde or benzyl alcohol as a carbon source and acted as a simple biocatalyst for which the authors develop a mathematical model. As expected, the reaction rate was related to the biomass concentration in the reactor, which fell slowly during the experiment due to escape from the system. No nutrients were provided in the input stream to allow regrowth of the cells. Unlike many experimental systems this was, therefore, a true cell reactor rather than a membrane fermenter. Fouling of the membrane would be expected to be less severe, due to the absence of new cell growth, enabling a relatively unsophisticated membrane unit to be used.

A novel use of a dialysis culture system is described by Marsot, Fournier & Blais (1981) to overcome the problems of culturing marine phytoplankton (*Phaeodactylum tricornutum*) in a continuous, high cell density, production system. The main role of the external dialysis unit, which was a standard medical haemodialysis unit, was to allow replacement of the sea-water nutrients and removal of waste metabolic products. They found the system advantageous, with low risk of contamination over prolonged periods. Cell densities of up to 4×10^7/ml were obtained.

There are several other situations in which membranes are used to retain cells as part of a functional system. In some cases, as in the construction of immobilized cell sensors, the cells or tissue slices form a thin layer over the surface of the probe and are held in place by the membrane. In other cases, such as the use of membrane inlet mass spectrometer probes, used to monitor volatile components in cell cultures, the membrane is used to exclude the cells from the probe. Weaver *et al.* (1980), for example, describe a small-scale system to hold

Saccharomyces cerevisiae temporarily in contact with an aqueous stream, which then passes over the inlet membranes of the mass spectrometer probe. Rather similar membrane inlets have been used by several other researchers. Most microbial fuel cells also have a membrane separator. Bennetto *et al.* (1983) used a cation exchange membrane, Nafion (Trade Mark of Du Pont Corp.) in a fuel cell containing *E. coli* or *S. cerevisiae* in the anodic compartment and buffered potassium ferricyanide solutions in the cathodic compartment. The use of appropriately charged membranes might prove beneficial in other membrane reactor systems.

A typical example of one type of microbial fuel cell was given by Bennetto *et al.* (1986). They used *Proteus vulgaris* in the anode compartment separated from the cathode compartment by an ion-exchange membrane. Thionine was used as the mediator and sucrose as the electron-donor substrate. Under appropriate conditions a quantitative conversion to carbon dioxide and electricity was achieved. The use of artificial membranes as carriers for immobilized enzymes and cells was reviewed by Gekas (1986), with particular attention to the membranes and the methods of immobilization.

As can be seen from this brief section there are almost unlimited possibilities to develop immobilized cell systems in which cell retention by membranes plays a significant role in the overall process, even where the membrane acts more as an ultrafiltration or dialysis unit rather than as a primary means of cell immobilization.

4   Covalent bonding, cross-linking and flocculation of cells

In theory, there is a clear distinction between cells covalently coupled to a support material and cells cross-linked to one another by covalent bonds. In practice, the situation is often more difficult to categorize because in some methods the cells may bind to one another as well as to the support. The support itself may also be formed by the cross-linking reaction, as occurs with proteins covalently linked with glutaraldehyde in the presence of cells. In the present chapter attention will be given to those situations where the mechanism is reasonably clear. It is also necessary to note that many support materials can adsorb cells without any pre-treatment, but may be activated (or as Kolot (1981) described it, 'grafted with coupling agents') for a more permanent immobilized cell preparation. There is a very extensive literature on the covalent bonding of previously isolated enzymes to activated supports and the interested reader is directed to that. In the present chapter only the covalent binding of cells is considered. In actual use there may be little difference, because one of the general problems with covalent bonding is that cells are exposed to potent reactive groups, which exert toxic effects. This is one problem which has restricted the use of the covalent binding of cells. The other major problem is the low cell loading which is achieved, compared to entrapment methods. Where viable cells can be covalently bound any cell division is likely to result in cell leakage from the support, because the daughter cells will probably be less strongly attached. Many industrial processes, however, involve the use of non-viable cells and are often restricted to single-enzyme conversions. There may be a considerable financial saving in immobilizing cells rather than extracting the enzyme and then immobilizing it. In addition, the direct coupling of enzymes to supports generally leads to a significant loss of activity. When the cells are covalently bound instead, the enzymes within them are protected against some of this inactivation.

From what has already been discussed in Chapter 1 it will be obvious that the outer surfaces of cells and cell walls contain large quantities of a variety of reactive groups, such as hydroxyl, aldehyde, ketone, carboxyl, amino, sulphydryl, imidazole and various substituted aromatic rings. There is a great potential, therefore, for the creation of covalent bonds with suitable activated carriers. The most commonly used coupling agent is probably glutaraldehyde although carbodiimine, isocyanate and amino-silane have also been used. A special category of coupling, described as partial covalent bond formation, occurs when hydroxides of titanium and zirconium are used.

An interesting approach to covalent binding was described by Marek *et al.* (1986). They activated the vicinal diol groups of the cell wall polysaccharides of two species of *Saccharomyces* by periodate oxidation and then linked the newly generated aldehyde groups to activated cellulose or glycidyl methacrylate. The enzymes catalase and glucose oxidase were also linked directly to the immobilized yeast cell walls The expressed cell-bound invertase activity was increased by the treatment. This method has many possible uses for biosensors and biocatalysis, due to the flexibility it offers in combining enzymic activity derived from different micro-organisms. Hu (1986) used epoxypolyamine as a multifunctional cross-linking agent for *Actinoplanes missouriensis*. This introduced a six-carbon spacer chain between the functional groups and gave good stability to the particles with a high glucose isomerase activity. It was suggested that the spacer effect might reduce diffusional limitation within the particles.

Very strong adsorption, as an essentially monolayer system, was described by D'Souza *et al.* (1986). They treated either a glass slide support or yeast cells with polyethyleneimine and obtained a preparation which retained cell viability. A simple, multiple slide reactor was used to show that no leakage of cells occurred, even with repeated use. Developments of this technique may be useful in biosensor systems.

The various categories of supports to which cells may be coupled have been classified by Kolot (1981). A first distinction is made between organic and inorganic supports. In the organic category are various natural polymers, which may also be used for entrapment, such as carrageenan, agar, alginate, pectins and cellulose. Various derivatives of natural organic polymers can also be used such as Sephadex^R (cross-linked dextrans), and carboxymethyl celluloses. The majority of these polymers are structural polymers from the walls of plant cells and can be easily linked to other cell walls by suitable bifunctional reagents. Proteins may also be used and again it is a structural protein, collagen, which is most commonly used. Among the purely synthetic polymers those used for entrapment, such as the polyacrylamides, may be used, as well as an almost infinite range of insoluble 'plastics', including phenolic resins and polystyrenes. The difference compared to entrapment is that the polymers are always pre-formed and act only as a support, not as a matrix in which the cells are trapped. It is therefore possible to use materials whose monomers, activators or polymerization conditions are, themselves, totally incompatible with the retention of enzymic activity.

Among the inorganic supports a wide range of natural and man-made materials is available. Alumina, zirconia, magnesia, silica, magnetite, diatomaceous earths, porous rock, coke, brick, ceramics and glass are all possible. Used without activation they are suitable supports for simple adsorption immobilization. With activation, they can be used to bind cells covalently with much less danger of leakage into the process stream. Whatever the support, it will clearly be the amount of surface area which determines the number of cells that can be bound. The conventional way of increasing surface area per weight of support is to use smaller particles. For some designs of reactors there is a limit (around 1 or 2 mm) below which the particle size becomes an operational problem. This problem is solved by using porous supports of an appropriate size and shape for the particular reactor being used. The size of the pores obviously varies with the type of cells being immobilized. The most thorough study of this factor is probably the work of Messing and coworkers at the Corning Glass Works, USA, mostly using porous glass. The pore size of many inorganic supports, however, cannot be varied at will during manufacture and a suitable material must be found by post-selection. In general, this is also a cheaper method, even if it is less intellectually satisfying.

Glutaraldehyde coupling

Polyphenyleneoxide is a porous polymeric support developed by Kubanek, Veruovic, Kralicek & Cimburek and covered by Czechoslovak Patent no. 1648-78. When used untreated its high surface area ($600 \, \text{m}^2/\text{g}$) allows cells to be easily adsorbed. The attachment is, however, very weak and cells of the yeast, *Kluyveromyces lactis* were completely lost after only one washing (Jirku *et al.*, 1980). The polymer readily reacts with glutaraldehyde in aqueous solution (optimum 5% v/v) and excess glutaraldehyde can be removed by washing with water. Cells added to the activated support polymer were taken up over a six hour period and approximately 70% of the initial activity (measured as β-galactosidase) was retained after 10 washing cycles. SEM studies revealed a relatively even distribution of the bound cells over the surface of the particles. In these experiments the cells had been previously dried to permeabilize them and they were probably not viable. Using essentially the same immobilization system Jirku *et al.* (1981) studied the production of steroid glycoalkaloids by cultured cells of the higher plant *Solanum aviculare*. The gel particles loaded with glutaraldehyde attached cells were used in a batch recycling column and extracellular liberation of glycoalkaloids shown to occur for 11 days. The technique was clearly promising, even for delicate plant cells, but does not appear to have been extensively exploited.

One way of improving the binding of cells to supports is to insert 'spacer' chains between the support material and the cell, a technique used routinely for enzyme immobilization. Jirku, Turkova &

Krumphanzl (1980) started with the hydroxyalkyl methacrylate gel Sepavon[R] H-1000 and reacted it with epichlorohydrin. This gel was then further reacted with tetramethylenediamine and finally activated with glutaraldehyde. After thorough washing and drying the activated polymer was allowed to take up a killer strain of *S. cerevisiae*. Over a 10 hour incubation the bound cell mass increased three-fold (on a dry weight basis), but only a few free cells were detected in the incubation medium and toxin production was demonstrated from the immobilized cells. SEM examination showed that the daughter cells had an abnormal, elongated shape, attributed to a continuing immobilization process. The possibility of making glutaraldehyde part of a mild, permanent, immobilization system was again shown by these results.

Almost any material can be surface-coated in such a way as to be capable of glutaraldehyde activation. One method, for example, is to coat with gelatin, which will then be both cross-linked to a permanent polymeric form and activated by glutaraldehyde treatment. The physical form (pellets, spheres, sheets, fibres), and whether solid or porous, does not have any influence, except by way of the area of reactive surface available.

Porous silica beads (Spherosil[R]), varying in diameter from 100 to 200 μm were used by Navarro & Durand (1977) to investigate the immobilization of the yeast *S. carlsbergensis*. Five different grades of porosity were used, having surface areas of 6, 34, 120, 180 and 445 m^2/g. The beads with larger pores, of course, have the smallest surface areas. The beads were first reacted with γ-aminopropyltrimethoxysilan and either used in this form or further activated with glutaraldehyde. Supports and cells were contacted together for one hour in a buffer at pH 5.0. Bound cell mass was determined by nitrogen content analysis and free cells remaining by Coulter Counter. Only 15 minutes' contact was found to be required for maximum cell immobilization using the highest porosity, glutaraldehyde-treated beads. Of the beads treated with the γ-aminopropyltrimethoxysilan only the lowest porosity (6 m^2/g) took up any cells. Of those additionally treated with glutaraldehyde the mass of cells immobilized increased with increasing surface area. Some metabolic distinctions were found between cells immobilized onto the glutaraldehyde-treated supports and those on the untreated 6 m^2/g surface area and between these and the cells attached to the 445 m^2/g support. Completely untreated silica beads were shown to induce similar metabolic changes to those treated with γ-aminopropyltrimethoxysilan. The question of the influence of pore size is returned to in Chapter 5.

Kim *et al.* (1982) compared the biotransformation of progesterone to 11-α-hydroxyprogesterone by mycelium of *Aspergillus phoenicis* immobilized in various ways. Albumen or gelatin was cross-linked with and to mycelial pellets, at very low temperatures ($-20\,^\circ$C) for several hours, before thawing and washing. The alternatives tested were alginate, carrageenan and urethane (PU-3 type) immobilization. In these experiments the rather harsh treatments involving glutaraldehyde resulted in very poor biotransformation. The paper is of additional interest due to the different ratios of the various possible products formed in the immobilization systems.

Carbodiimide coupling

Kolot (1981) reported on the activation of carboxymethylcellulose with carbodiimide and its use to immobilize *Micrococcus luteus*. The binding takes place in the presence of *O*-acylisourea. The cells retained the target enzyme activity of histidine ammonia lyase. Comparison showed that modified celluloses with longer 'spacer' chains were better for immobilization. This was explained by the greater availability of reactive sites for cell binding. Other researchers have found that carbodiimide coupling was highly toxic and resulted in the loss of both viability and biocatalytic activity of several different bacteria.

Metal hydroxide coupling

Many metals are capable of forming gelatinous precipitates of their hydrous oxides. These are commonly called hydroxides and those involving titanium or zirconium have proved of value as a mild, simple method of cell immobilization. They are insoluble in water over the pH range normally suitable for cells. Kennedy (1979) discussed these methods in some detail, suggesting that hydroxyl groups on the surface of the precipitates are replaced by a possible variety of groups, from amino acids in proteins or from carbohydrates. The hydroxides were formed by neutralizing an acid solution of the chlorides of the respective metals. The precipitates were washed with saline and the cell preparation mixed in and allowed to stand. If lower pH values are required by the cells the use of titanium (IV) hydrous oxides is recommended, allowing the immobilization of *Acetobacter* species. Other species bound were *S. cerevisiae, E. coli* and *Serratia marcescens*. Cells were capable of respiration, but at a lower rate than free cells. There was negligible leakage of *S. marcescens*. The use of *Acetobacter* species in a tower fermenter, for the production of acetic acid, was investigated. The immobilized preparation gave a three-fold increase in output per day.

An extension of this technique was investigated by Dias, Novais & Cabral (1982). Two inorganic supports were used: pumice stone and Spherosil^R XOB-015 a porous silica. They were activated with titanium chloride solution for 48 hours. Comparable samples of titanium-activated Spherosil^R were treated with hexamethylenediamine to produce the alkylamine derivative. Further activation of some samples was carried out using glutaraldehyde. In each case *S. cerevisiae* cells were allowed to immobilize onto the supports over a two hour period. Initial invertase activity (the target enzyme) was least on the glutaraldehyde-treated support, almost two-fold higher on the titanium hydrous oxide alone and four-fold higher on the alkylamine. After a day or two the higher activities had fallen sharply and from that time onwards, there was little difference between the treatments. The simple titanium hydrous oxide linking gave the best stability in long-term batch tests. The macroporous pumice stone had a greater capacity than the Spherosil[R], yielding an activity superior to alginate-entrapped cells. There would appear to be many possible uses of the mild metal hydrous oxide system but considerable further work is needed.

The continuous production of acetic acid by *Acetobacter* species, immobilized onto hydrous titanium-chelated cellulose and hydrous titanium oxides was investigated by Kennedy *et al.* (1980). Chromatographic grade cellulose powder (Whatman[R] CF11) was reacted with titanium chloride in acid solution for two hours, then dried and washed. Two strains of *Acetobacter* were used, one producing an extracellular slime and showing slight self-aggregation, while the other showed no self-aggregation ability at all. The aggregation strain showed a visible accumulation of biomass particles in the presence of titanium hydrous oxides over a 10 day run in a fermenter and gave higher acetic acid production in long-term continuous operation. With the non-aggregation strain no effect was seen from hydrous titanium oxides alone, but a significant increase (of up to 20%) was found with the cellulose-titanium hydrous oxides addition. The aggregating bacterial strain is actually a mixed culture and the slime contained cellulose produced by *Acetobacter xylinum*. The authors discuss the mechanisms of immobilization in some detail in relation to improved continuous acetic acid production. This particular application of the hydrous titanium oxides method clearly falls somewhere between immobilization onto an activated support and a form of assisted 'nucleated' flocculation.

Flocculation

Flocculation is the aggregation together of individual cells into large clumps which, because of density or other altered physical properties, usually tend to settle out from free suspension. Liquid or gas flow, or gas produced by the cells themselves, will tend to keep the flocs in suspension. Flocculation may be a natural feature of certain species or even of only certain strains within a species. It may occur throughout the whole lifetime of a culture or spontaneously at the later stages of a fermentation. In other cases cultural conditions or the addition of specific agents may initiate flocculation. The flocs themselves may be unstable and easily disrupted or strongly coherent and capable of being easily handled without falling apart. There has been an enormous amount of research in relation to flocculation. This has primarily been directed towards the study of yeasts in the production of alcoholic beverages and, more recently, industrial alcohol. A second major area of interest has been the settling out of excess biomass in the wastewater treatment industry. If flocculation can be controlled, both in species where it is a natural phenomenon and in other species, it provides a simple, gentle, effective and cheap means of cell immobilization.

The mechanism of natural flocculation has been a subject of much discussion. In the case of yeasts, a major role has been ascribed to the interaction between the negative surface charge on the cell wall and the presence of cations in solution. It is a matter of long-standing observation that strains of yeasts differ in their capacity to flocculate and that this is an inherited characteristic. A mixed population of cells can be easily separated into flocculating and non-flocculating forms. This is achieved by simple sedimentation from a batch culture, or by using either a separator tank or self-selection in a continuous culture system. In the latter case the dilution rate is adjusted, in an appropriate design of fermenter (generally of 'tower' design) so that individual free cells are washed out and the flocculent masses remain in the fermenter. Prince & Barford (1982a) took a definitively non-flocculent strain of *S. cerevisiae* and grew it in an up-flow tower fermenter provided with a baffled settling zone around the overflow weir. Starting with a low value (0.06) the dilution rate was incrementally increased after aggregation was observed. In its original form the yeast rarely produced clumps of even three or four cells. It took 72 hours for the first small flocs to appear, followed by a rapid rise in cell density in the fermenter. After 170 hours the dilution rate was increased to 0.11. After further increments the dilution rate could be brought to 0.6, at which point excessive substrate carry-over commenced, although floc retention was still reasonably good. This work illustrates clearly the relative ease of selecting flocculent forms, even from strains previously characterized as non-flocculating, given sufficient time and a suitable selection system.

Hsiao *et al.* (1983) took the fission yeast, *Schizosaccharomyces pombe*, and used a batch flask system, with decantation of spent medium, to obtain a flocculent form. Variations in the glucose content of the medium, from 10 to 100 g/l did not significantly affect flocculation. Although growth was quite good in a medium with 10 g/l glucose and 'yeast nitrogen base' alone, it did not produce good flocculation. Addition of yeast extract to the medium gave extensive flocculation but bactopeptone in the medium was inhibitory to flocculation. Complex interactions obviously occurred. Growth in a continuous column fermenter for even a short produced large flocs (approximately 1–3 mm diameter) compared with the small flakes (50–200 mesh size) obtained in batch flask growth. Flow rate influenced floc size, too little allowing flocs to collapse and settle as a dense mass. Too high a flow rate caused wash-out of the flocs. Clearly, the selection of suitable conditions for flocculation may not always be as simple as the work of Prince & Barford (1982a) suggests, but the design features of the fermenter and the selection system are probably the major factors responsible for the success of the latter authors.

Ramirez & Boudarel (1983), among many others, have emphasized the crucial role of the cell separation stage for cell recycle systems in their studies of ethanol formation from beet juice. In this type of application it is likely that only gravity settling will prove economic. One approach to improving this is to add agents which will promote flocculation by acting as dense nucleating centres. This might be considered as a form of adsorption onto extremely finely divided support material but in actual use is sufficiently distinct to be seen as an extension of flocculation. Weeks, Munro & Spedding (1982, 1983) used inert powders to promote flocculation in a non-flocculating strain of *S. cerevisiae*. The agents tested were nickel powder (3–7 μm), milled iron sand concentrate (to pass a 37 μm sieve), technical grade iron oxide powder and technical grade calcium carbonate. In preliminary experiments, settling rates were determined in distilled water containing the yeast cells. Iron oxide and calcium carbonate were required in smaller quantities than the nickel powder and considerably more of the iron sand was needed, probably due to its much greater particle size. The nickel powder gave rapid settling, had a small, uniform particle size and was stable under acid conditions. It was selected for further study, although it was suggested that iron sand, being cheaper, might be more suitable for industrial use, if it could be ground finely enough. Seeding increased the density difference between flocs and liquid by just over a 10-fold factor and increased particle size, both leading to more rapid settling.

Moving to experiments on the influence of media components, it was found that yeast extract and citrate buffer strongly inhibited flocculation. Although the pH value in the range 2.0–9.0 had little effect on flocculation, it was found that a rapid change in pH from 4.5 to 8.0, with some stirring, rapidly produced large flocs, even in the presence of media components. A similar effect on natural flocculation of other yeasts has been previously noticed. Surface loading of the nickel particles, as seen with the SEM, was much higher with the pH switching technique. Extremes of either acidic or alkaline pH values, of course, produce lysis of the cells. The authors conclude their second paper Weeks *et al.* (1983) with an extensive discussion of flocculation. They consider the special behaviour of their system to be due to the formation of a pseudo-equilibrium. The yeast cells start off with a negative surface charge and the nickel particles with a positive surface charge. Resultant electrostatic attraction leads to small flocs, but the charge repulsion between these prevents larger aggregates forming. When the pH is rapidly changed from acidic to alkaline, by the addition of alkali, some of the positive surface charge of the nickel particles would be reacted with the hydroxyl ions. The resulting subtle and patchy (due to incomplete mixing) surface charge changes would result in some net positively-charged aggregates and some net negatively-charged particles, capable of aggregation to the larger floc sizes. The exact conditions are clearly important if this pseudo-equilibrium state is to be achieved. It is, however, a reproducible system, with many possible applications. Reversibility of flocculation allows maximum dispersion of active cells in the reactor, coupled with easy cell recovery and recycle. Used in the way proposed, however, it would be expected to have little effect upon the half-life of the biocatalyst particles and essentially consists of a fermenter rather than an immobilized cell reactor system.

The simple use of flocculent strains of yeast forms part of many beer-making processes, as a low-cost means of removing yeast biomass from the product. The use of flocculent strains during continuous alcohol production is often associated with the use of tower fermenters allowing floc retention. Typical of research in this area is the paper by Prince & Barford (1982b) using *S. cerevisiae* and *S. diastaticus*. At high dilution rates these approach a fluidized-bed mode of operation, due to gas evolution and liquid flow. High throughputs of medium did not appear to alter floc characteristics and the limit to dilution rate increase was the extent of conversion of substrate achieved rather than yeast wash-out. The same authors (Prince & Barford, 1982c) used an essentially similar tower fermenter to study ethanol production by *Z. mobilis*. Over a 10 day start-up period at a low dilution rate (0.065) a flocculent form was selected out. This is somewhat longer than the time needed for yeast in their 1982a paper (previously mentioned above). No attempts were described to optimize long-term, steady-state performance, but the feasibility of the selection and operational techniques was clearly demonstrated. The flocculation approach was considered to have advantages over various other immobilization methods discussed in the introduction to their 1982c paper.

Arcuri, Worden & Shumate (1980) had previously compared a non-flocculating strain of *Z. mobilis*, adsorbed into 2 mm × 2 mm pads of glass fibre, with the same strain grown in the presence of polystyrene beads (1 mm diameter). In the latter case the authors found self-flocculation of the bacteria, but no significant attachment to the polystyrene beads themselves. In the case of the glass fibre pads the growth was clearly retained in the pads and may have been actually attached to the fibres. Their system, with its requirement for a support material used in a column reactor under fixed or expanded bed conditions, is not as simple as the selection of flocculent strains of *Z. mobilis*. Strandberg, Donaldson & Arcuri (1982) compared the performance of a flocculent strain, available from a culture collection, in three different types of bioreactors. These were: a simple upright column; an upright column, tapered at the base, with a short, angled side arm; and a column held at an angle of 45 ° from the vertical, with several gas vents along its upper side. The best design was the upright tube with the angled side arm, which allowed liquid outflow to occur with minimal floc wash-out. This design has some functional similarities to the tower fermenter with baffle-protected outflow, as used by Prince & Barford (1982c).

Lee, Skotnicki & Rogers (1982) considered the kinetic behaviour of a mutant flocculent strain of *Z. mobilis* produced by treatment of a non-flocculating strain with nitrosoguanidine. Compared with the parental strain, the flocculent form had a slightly reduced (by 11%) specific growth rate and showed a 23% reduction in the specific glucose uptake rate and a 22% reduction in ethanol production rate. These were attributed to diffusional resistance in the flocs. It is likely that similar diffusional restriction occurs with all flocculent strains of micro-organisms. This is generally considered to be more than compensated for by improved handling and operational characteristics but could be taken as a point in favour of reversible induction of flocculation, as needed, for cell recycle or product stream clarification. In their paper Lee, Skotnicki & Rogers (1982) referred to unpublished information elucidating the mechanism of flocculation in *Z. mobilis*, but gave no details. Fein *et al.* (1983) refer to their own work suggesting that thin fibrils (possibly cellulosic in nature) are responsible for holding the cells together. They used a simple single-stage continuous stirred-tank reactor (fermenter) for their studies on ethanol production, using a conventional overflow weir. No external settling system was used, but a slow nitrogen sparge stream was used to assist carbon dioxide removal. The authors suggest that scale-up may be easier on their CSTR system as compared to tower fermenter systems, which have given similar performance data on flocculent strains of *Z. mobilis*. The selection of flocculent strains of *Z. mobilis* in continuous conical reactors, at high dilution rates, is a reliable method of improving the concentration of active biomass for ethanol production (see, for example, Baratti, Varma & Bu'Lock, 1986).

The literature on the flocculation and settling characteristics of both aerobic and anaerobic sludges in the wastewater industry is very extensive and it is only appropriate in the present book to mention this subject briefly. Flocculation is an essential requirement of the active mixed-culture biomass in the activated (aerobic) sludge process. The material forming the flocs both acts as an absorbant for heavy metals and other pollutants and also promotes the rapid settling-out of the flocs, after the treated effluent has passed out of the digester into the settling tanks, before discharge from the plant. This separation is essential to provide active biomass for recycling to the digester and to prevent the discharge of particulate organic matter, which would itself be a polluting agent. Clearly, there is an element of self-selection for flocculating characteristics, simply because of the operating regimes, but there may still be problems from time to time and much work has been directed towards elucidating and overcoming these difficulties. Thompson & Forster (1983) compared three bacterial strains, one flocculating, one non-flocculating form of the same species and one filamentous. It is generally accepted that extracellular polysaccharides are important in flocculation and these were studied under various test conditions. Rather surprisingly they found greater carbohydrate produced when the C:N ratio was 10:1 than when it was 25:1. It is generally assumed that, under conditions of excess carbon and nitrogen deficit, more extracellular carbohydrate will be formed, so the expected result would be the other way round. The dissolved oxygen level had a significant effect, with less carbohydrate produced at the lower level tested. The hexuronic acid content of the carbohydrate contributes to the surface charge of the floc particles. For the filamentous and flocculating bacteria the hexuronic acid:carbohydrate ratios were almost identical and there was a lower hexuronic acid content (and hence surface charge density) at the lower of the two dissolved oxygen levels tested. The authors found that comparison of the results between the flocculating and non-flocculating bacteria of the same species indicated that the amounts of hexuronic acids present were probably related to their respective flocculating capabilities. Clarke & Forster (1982) had earlier shown that the amounts of extracellular polymers extracted from activated sludge could be correlated with operational characteristics of the process from which the original samples were taken.

Similar attention has been paid to anaerobic digesters with regard to flocculation characteristics. For efficient operation of the upflow anaerobic sludge blanket (UASB) reactor, for example, a well-adapted flocculent (or pelletized as it is generally described in this application) active biomass is required. At present this is not always possible to achieve under all feed and start-up conditions, but the UASB reactor, like the tower fermenter systems discussed for yeast, operates in such a way as to select for flocculent forms during continuous operation. It is also probable that fine sand and silt particles act as nucleating agents for such flocs. Although flocculating agents are frequently added at the settling tank stage, to improve sedimentation, they are less often used in the digester itself. Callander & Barford (1983) used the synthetic flocculating agent Zetag^R 88N (selected from a wide range originally tested) as being non-toxic, non-degradable and effective. They compared the performance of two anaerobic stirred-tank digesters for waste cheese whey treatment. The one with flocculating agent present had over twice the concentration of active biomass and could cope with a 50% greater organic loading than the control digester. Conversion efficiencies were similar in both. The cost of the flocculent agent was only a few per cent of the value of the methane produced and was considered to give a significant improvement in performance.

Moving to another application of flocculating agents, Nazly & Knowles (1981) examined the immobilization of mycelia of *Stemphylium loti*, in which cyanide hydratase had previously been induced. The purpose was to develop a cyanide detoxification system. They used the polyelectrolyte flocculating agents A10[R] and C7[R]. The immobilized mycelia retained 55% of the target enzyme activity initially present after three days at 24 °C, while non-immobilized mycelia had completely lost their activity after 16 hours. The apparent K_m of the enzyme in the immobilized mycelia was 43 mM compared with 27 mM for control mycelia. In experiments with the immobilized mycelia in a column reactor, glucose addition to the feed extended the operational period for cyanide detoxification. For this particular application alginate cannot be used as an immobilization matrix because of the ability of cyanide to complex cations. Stability of K-carrageenan and polyacrylamide immobilized mycelia was not as good as that of flocculated mycelia. It would appear that many other possible applications of flocculating agents exist. This simple, cheap form of immobilization deserves much greater study than the rather restricted fields of ethanol production and wastewater treatment to which it has mostly been applied.

5   Adsorption

Immobilization by adsorption is a process involving surface interactions between cells and support materials. The forces involved are mostly those of ionic and hydrogen bonding which, although being individually weak when compared to covalent bonds, are nevertheless capable of producing relatively firm binding if the number of bonds is sufficiently large. Because only the natural properties of the surface and cells are involved, with no toxic influences arising from activating reagents, the method is very mild, provoking little damage to cellular metabolism or loss of viability. The method is relatively non-specific, although the type and condition of the cells often exert an important influence. Although cells often adsorb to surfaces easily, they may also desorb back into the free suspended cell state. This leakage may cause serious problems in an immobilized cell reactor, due to loss of active biomass and contamination of the process stream. Desorption may be promoted by many factors. Changes in ionic strength of the pH value of the medium may cause severe sloughing of the cells. Continued growth of adsorbed cells may exceed the loading capacity of the support material. The development of moribund regions close to the support surface, due to diffusional limitations on nutrient, product or gaseous exchange, may cause disintegration of the active cell layer. Gas or liquid shearing forces, caused by operating at too high an aeration or dilution rate and the scouring effect of particulate material in the process feed stream may also cause loss of biocatalytically active cells. If the desorption process is controlled it is possible to use it beneficially to replace spent (non-active) biomass by new cells of the same or a different type.

Immobilization by self-adsorption onto surfaces is a feature seen in many natural habitats, particularly those where a large bulk flow of liquid occurs. The operation of conventional wastewater treatment plants (of which sewage works form one important type) involves the deliberate encouragement of microbial growth on filter beds or other supports. In this case the choice of material has most often been based upon tradition and simple economic factors, with little real attention to the mechanism or subtleties of adsorption. Recent attempts to improve the efficiency of such processes and reduce the size of treatment plants have resulted in many new developments. Other researchers have focused their attention onto the problems of biofouling, for example in metal pipes under a

135

variety of process conditions. It is not the purpose of this book to review these substantial topics although reference may be made to them from time to time. The interested reader should consult appropriate texts.

There are clearly two sides to adsorption which can be investigated, both independently and as a complete system. The first relates to the principles which can be applied in the selection of support materials and the second to any measurable features of cells which can be correlated with adsorption. What are the actual processes occurring during adsorption? In her review of microbial carriers, Kolot (1981) first draws attention to certain primary features. Supports must be non-toxic to the cells being used. This does not mean that the supports must necessarily be inert, because many surface reactions occur during the adsorption of cells. Because the significant feature of any immobilization process is the amount of active biomass present, it clearly follows that the cell retention capacity of the support must be high. Surface area alone is not a reliable guide to the amount of biomass which can be retained, because accessibility of that surface is also essential. Porous supports must therefore have pores of appropriate dimensions. Compared with entrapment there is no doubt that simple adsorption generally results in much lower cell loadings per gram of support material used. In a large number of applications it is essential to maintain a pure culture system and this requires the sterilization of support material before loading of the selected cells. Because heat sterilization is by far the simplest, safest and generally used method, it is highly desirable for support materials to be capable of steam sterilization. Stability under the operating conditions, especially with regard to pH value and ionic strength is obviously a necessity. The mechanical properties of the support and its availability in various physical shapes become of major importance in the selection of an appropriate cell reactor. In general, approximately spherical particles allow the widest choice of reactor designs and operational regimes. Finally, the availability, ease of preparation at an appropriate production scale and the cost of the support need to be considered. Although reusability has been suggested as a desirable characteristic, it is generally of no significance except where the support material is expensive and would therefore justify recovery costs.

Atkinson *et al.* (1980) considered the properties of desirable support systems for the adsorption of micro-organisms, with particular reference to the practical aspects of process intensification. Some desirable features of support materials are given in Table 5.1. Apart from comparative aspects of reactor systems (which are discussed in Chapter 7) they consider the potential and limitations of flocs and films, leading to a consideration of the desirability of accumulating active biomass in pre-fabricated support particles. These support particles have high void spaces inside them, which allow biomass to accumulate under low shear

Table 5.1. *Some desirable features of supports for cell adsorption*

Non-toxic material
High cell retention capacity
High cell loading capacity
Stable to heat sterilization
Stable at appropriate operational pH values
Resistant to microbial degradation
Availability in appropriate shapes and sizes
Cost appropriate to the application
Density appropriate to reactor type used
Reusable

conditions while the liquid phase outside the particles can be moving at a speed appropriate for efficient nutrient, product and gas exchange. Because of abrasion between particles, any excess biomass accumulation over the surface of the support particles can be removed by movement during reactor operation. In this paper the biomass support particles are made of stainless steel mesh formed into a spherical shape. Stainless steel fibre sheets were used by Juneja *et al.* (1986) in a laboratory-scale three-stage horizontal parallel flow reactor system. The sheets were used vertically, in units of 26 sheets, for the continuous production of ethanol. A flocculent strain of *S. cerevisiae* was initially recycled through the system to charge the fibre sheets. When required, the units could be cleaned for reuse by placing them in boiling water. The flow pattern in the reactors was essentially plug-flow, which is considered to be the best type of reactor for product-inhibited processes.

Other support materials can be used in a similar manner. In their 10-fold list of beneficial features Atkinson *et al.* (1980) include the following points. The particles allow continuous use, under a variety of growth conditions, without adverse effects on conversion efficiency. Almost any organism can be retained in the particles. The particles can be of any chosen size or shape appropriate to the reactor design being used. The support particles are not disrupted by gas production. High active biomass loadings (up to 25 g dry weight per litre of bed volume) can normally be obtained in the reactor. The particular substrate/micro-organism system being used has little influence upon the concentration of active biomass which can be achieved or the conversion efficiency of the reactor. Biomass can generally be easily removed from the support particles so that they can be reused. Where filamentous organisms are being used the biomass is retained within the support particles and the usual problems of increasing liquid-based viscosity and flow obstruction by hyphal mats are avoided. By controlling the number of support particles in the reactor it is possible to fix the overall concentration of active biomass at any pre-set level in the reactor. By appropriate selection of the support material, particles of any required density can be produced to suit the operational requirements of the reactor system. If required, the porosity of the particles and the material they are constructed from can be selected for particular applications.

At the opposite end of the pore size investigations, the researches of Merssing and coworkers are considered classical. In Messing, Oppermann & Kolot (1979) for example, the results are given for adsorption of several different micro-organisms into porous carriers, mostly of fritted glass. The pore sizes ranged from 1.1–195 μm in average diameter and this was related to the size of the micro-organism. Studies included *Escherichia coli, Serratia marcescens, Bacillus subtilis, Saccharomyces cerevisiae* and *S. amurcae,* and *Aspergillus niger* spores. The adsorbed cell mass was quantitatively determined by extraction of the cellular ATP and its measurement by bioluminescence using the standard luciferin–luciferase system and a Du Pont Biometer[R]. A standard protein determination was used to quantify the uptake of spores of *Streptomyces olivochromogenes* and *Penicillium chrysogenum.* For those species immobilized in spore form, growth was also allowed to develop in the pores by incubation in an appropriate medium. For some organisms coupling agents were used to ensure adequate 'adsorption' into the pores, but this does not influence the interpretations of the results of the influence of pore size now being considered. For the three bacterial species reproducing by simple fission, it was concluded that maximum accumulation of biomass occurs with pores of from one to five times the bacterial cell size. The figure cannot be more precisely defined due to unavoidable variations both in cell size, which cannot be prevented under normal culture conditions, and in pore size, which cannot be restricted to a single diameter during support preparation. With the two yeast species which reproduce by budding, a rather broad range of optimum pore size was recorded, with evidence of two peaks in it, probably corresponding to single and budding cells respectively. Pores between the smallest diameter up to three to four times the largest diameter were indicated. With *A. niger* spores were immobilized best in pores approximately the average spore size but, as expected, larger pores (to 16 times the largest spore size) are needed to accommodate subsequent mycelial growth. With *S. olivochromogenes* an influence of surface charge was noted for initial accumulation and growth, but became insignificant after substantial growth had occurred. A value of 16 times the largest spore size gave a high biomass accumulation. This same 16 times relationship was concluded for the rather complicated results obtained with *P. chrysogenum.* The theoretical value of these results is great but, unfortunately, few process developments appear to have made any significant use of this information, nor, indeed, to have used the rather expensive controlled-pore-size fritted glass supports.

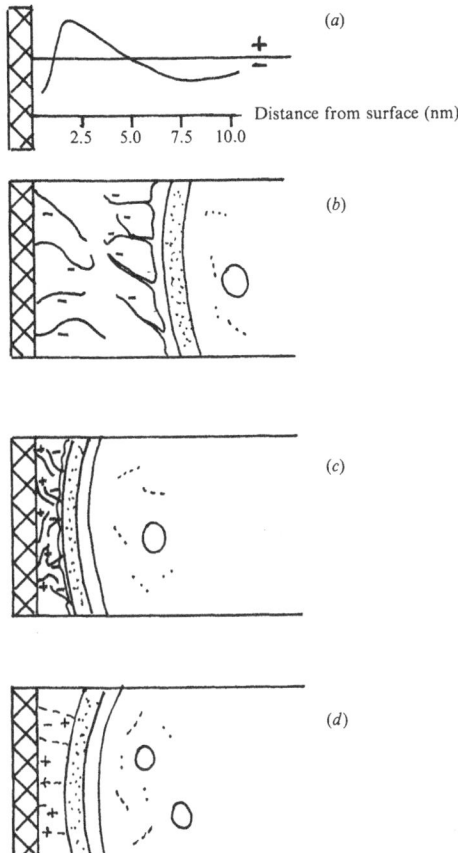

Figure 5.1. Diagrammatic representation of the initial stages in the adsorption of a cell to a support surface. (*a*) Energy difference between attractive and repulsive forces at various distances from a support surface; (*b*) steric and charge repulsion but with narrow processes that may penetrate the energy barrier; (*c*) polymer bridging between the cell and material adsorbed onto the support; (*d*) additional short-range ionic and hydrogen bonding, particularly important to cells without much extracellular polymeric material.

Turning to the factors involved in cellular adsorption, some of the key concepts are developed in an excellent paper by Gerson & Zajic (1979). The mechanisms of attachment of cells to surfaces may involve electrostatic attraction, extracellular polymeric products of the cells and projections sticking out from the cell surface (Figure 5.1). Excluding the last situation, which is related to the special properties of individual types of cell, attention must focus firstly on the electrostatic forces and then on the surface free energies of the support material and the cell.

Electrostatic forces are highly significant in the initial stages of adsorption and any electrostatic repulsion which exists must be overcome before a cell can be immobilized onto a support material. As was discussed in Chapter 1, the outer surface of a cell is generally charged and in most cases also contains a variety of ionizable groups. Any cell will normally have a rather diffuse 'cloud' or 'shell' of ions around it and the net charge will be measurable as the zeta potential of the cell surface. Support materials also have zeta potentials. If cell and support have zeta potentials of opposite signs then there will be a motive force for attraction, leading to immobilization at the interface. As discussed later in this chapter, this can be put to positive benefit by using ion-exchange materials as supports. The existence of similar charges on both cell and support does not necessarily prevent immobilization but the repulsive forces between the two must, in some way, be overcome, perhaps only in certain places by, for example, surface projections. Michaux *et al.* (1982) considered ways of altering the surface charge of cells and supports by coating them with various macromolecules. It is clear that if a negatively-charged cell is exposed to a positively-charged polymer, or vice versa, the surface properties of the cell can be completely changed by the masking polymer. Dilute solutions of divalent salts (such as calcium chloride or aluminium sulphate) have been shown to abolish zeta potentials on cell surfaces and may also act as bridges between two negatively-charged surfaces. Michaux *et al.* used yeast cells of three species (*S. cerevisiae*, *Candida pseudotropicalis* and *Kluyveromyces fragilis*) and sawdust as the organic support. Addition of gelatin and cationic starch produced positive zeta potentials in place of the original negative ones for all three species. Inulin and carrageenan produced more negative potentials for *C. pseudotropicalis* and *K. fragilis* but had negligible effects on *S. cerevisiae*. Protease peptone had little effect on the three species, making the zeta potential slightly less negative, in all cases, when used at up to 500 μg/ml. When zeta potentials of *S. cerevisiae* were measured in culture medium, instead of the distilled water used to obtain the previously quoted results, it was found firstly that the normal cell zeta potential was of the order of -1 to -7 mV. In distilled water the value was between -36 and -50 mV. In addition, cationic starch caused only a slight shift to zeta potentials, of approximately $+4$ mV. In distilled water the value after addition of the same quantity of cationic starch was $+6$ mV. In distilled water sawdust had a zeta potential of -40 mV. Cationic starch at 100 ng/ml converted this to $+6$ mV. The concentration of soluble organic polymer used was critical in determining the actual values of zeta potentials for cells and sawdust. By careful selection of the polymer and its concentration it was possible to produce cells and sawdust of opposite charge signs. In a cell reactor the presence of cationic starch (at 5 ppm) was shown to reduce cell desorption from the sawdust from 50% to 27%. Clearly, a consideration of the surface charge of a support and the manipulation of the zeta potentials of cell and support may be a fruitful avenue of research. It is, however, true to say that most, if not all, of the systems actually used have been found by entirely empirical experiments rather than by a reasoned consideration of zeta potentials.

Turning now to the matter of surface free energies the situation is rather similar. To measure the actual value of the surface free energy poses some serious problems and the most generally accepted method is to determine the critical surface tension, which is that at which a liquid will spread out over the surface rather than form a discrete droplet. Test liquids with known surface tensions are available but some problems arise when using polar liquids. Most of the commonly used plastic supports have surface energies of the order of 30–46 ergs/cm^2 (Gerson & Zajic, 1979). The presence of water lowers the surface energies, as is dramatically illustrated for glass. In the dry state a value of 700 ergs/cm^2 is found. At 0.6% relative humidity this drops to 45 and at 95% relative humidity to approximately 31 ergs/cm^2. The difference in energies between the surface free energies of the interfaces before and after separation gives the free energy of adhesion. Calculated values of this appear to be considerably different (often by a factor of 100 or 1000) from those obtained by direct experiment, probably due to the existence of weaker fault regions in real, as opposed to theoretical, joints. In the case of cell adhesion, however, the joints may more closely resemble the 'perfect' state, due to their small size and the presence of space-filling polymeric exudates. As with a glued joint in woodworking, a good adhesion confers the ability to resist tensional and shear forces, of which the latter is most relevant to conditions in an immobilized cell reactor. The intermolecular forces involved in cell binding consist of those associated with hydrogen bonds, with a bond energy of around 21 kJ/mole of bonds and the weaker Van de Waals forces. Both of these, as well as the stronger covalent bonds, play a role in keeping extracellular polymeric exudates around micro-organisms, as well as in attaching them to support materials. It is possible to have a system in which the cells, under the influence of shear forces, might separate from their own polymeric exudates, rather than the cell plus its exudates separating from the immobilization support material.

The measurement of the surface free energies of cells is even more difficult than that of support materials. In the most common method the contact angle is measured between normal saline and the surface of a film of the dried micro-organism. Obviously, changes will occur in the cells during the preparation of the film. The surface energies for a range of cell types fell mostly in the range 63–68 ergs/cm^2 (Van Oss, Gillman & Newmann, 1975). Organisms specially evolved for particular habitats may fall well outside this range. *Corynebacterium lepus*, for example, which uses solid and liquid alkanes, has a value of 48.7 ergs/cm^2 while *Thiobacillus thiooxidans*, attaching readily to elemental sulphur, has a value of at least 72 ergs/cm^2. Plots of the relationship between cell adhesion and the surface free energies of solid supports give a graph which indicates a maximum around the point where the presumed surface free energies of the cell are similar to those of the substrate. In general, cells adhere easily in large numbers to low surface energy supports but the attachment is weak and the cells are easily removed. High surface energy supports may only slowly accumulate a few cells, but these are likely to be relatively firmly attached. Working with *Pseudomonas* NCMB 2021, Fletcher (1976), showed that adhesion to polystyrene follows the Langmuir adsorption isotherm. The value of K, the adsorption coefficient, was of the order of 5.0×10^8 but was variable depending on the age of the culture. It is well known that the surface properties of cells differ with the age, cultural conditions and often the composition of the culture medium. Results from other researchers show clear difference between species. As with the use of zeta potentials, however, the gap between the empirical selection of support materials and the more elegant practical application of theories of cell adhesion remains very large.

The accumulation of nutrients, as well as micro-organisms, at the interface between a solid support and a liquid is well known and the two events are undoubtedly related. Three-phase interfaces, between solids, liquids and gases, also occur and may accumulate even higher concentrations of nutrients and cells. There has been much research carried out by microbiologists dealing with the relationship of micro-organisms to surfaces. The phenomenon of surface-associated growth is discussed by Ellwood *et al.* (1982). Under natural conditions and in complex fermentation media, any exposed surfaces rapidly become coated with polymeric and other substances from the bulk liquid phase. For this reason, short-term studies on the attachment of micro-organisms to new clean surfaces may give a rather abnormal view of the adsorption phenomenon. The important influence this may have is obvious from the previous paragraphs. The more permanent attachment of micro-organisms to surfaces generally involves extracellular products of the micro-organisms themselves and it is for this reason that some species are capable of producing stable films while others (or even different strains of the same species) may adsorb but also easily desorb.

The importance of the physiological state of a cell in influencing its adsorption onto a support was clearly demonstrated by Bringi & Dale (1985). They showed that nutrient starvation of *S. cerevisiae* enhanced self-adsorption into glass fibre mat supports of 200 μm median pore diameter. Cell loadings of 600 mg dry cells/g support were obtained. On supports grafted with gelatin and treated with glutaraldehyde before use, the resistance to shear was increased by about 50% when cells were starved in a sodium chloride containing medium. Starvation in distilled water reduced adhesion.

It is worth noting that conventional liquid culture systems, especially stirred fermenters, often lead to the replacement of strains producing extracellular polymers by those which do not. There may be some point, therefore, in going back to original isolates and relatively gentle culture conditions if the aim is to produce a large bulk of organisms for immobilization by adsorption. With entrapment or covalent linking this would be of little relevance. The third aspect of a successful adsorption system of immobilization is the increase of active biomass brought about by growth. Too little, and the biocatalytic activity will be too low. Too much growth may result in either surface films with severe diffusional restrictions or leakage of the excess biomass. The natural exopolymers of the attached organisms will have an important influence on both of these, but it should be remembered that the properties of the microbial film will be significantly influenced by the shear forces developed by liquid and/or gas flow over the surface. With moving spherical particles, such as those promoted by Atkinson *et al.* (1980) it is not clear whether the inter-particle surface abrasion slows growth or merely provokes continuous cellular damage, causing the release of fragments of damaged cells or even fully viable whole cells. No doubt this will depend very much upon the cell system being immobilized. For non-biological adsorption, the so-called DLVO (after Derjaguin & Landau and Vernwey & Overbeek) theory provides an acceptable mathematical treatment of colloid stability but it is not considered to be particularly useful in the discussion of films of micro-organisms, because it is impossible to obtain accurate values for several of the key parameters.

In general, the literature on microbial adhesion to surfaces can be divided into two distinct areas. Firstly, there is a wealth of highly detailed information on the attachment of viruses and bacteria to animal cell surfaces and body membranes. This is of marginal interest to the biotechnologist, unless he is specifically dealing with such organisms. A rather more slowly developing literature on the attachment of phytopathogenic and symbiotic micro-organisms to plant cell surfaces is also available. The literature of most relevance to biotechnological aspects of cell immobilization is obviously that which deals with the attachment to non-living organic and inorganic surfaces. Several workers have reported on the essential need for 'inert' surface supports to be present if growth is to occur in very dilute nutrient media. In practical terms this may mean that adsorption-immobilized viable micro-organisms may be capable of growth in much more dilute nutrient media than when free-cultured. This ability to survive at below the normal fermenter 'maintenance' levels of nutrients may have a significant influence on the extended life-span shown by immobilized cells. This aspect appears to have been little studied by biotechnologists, who have generally opted for the alternative of rapid regrowth under high nutrient concentrations to replace failing biocatalytic activity.

Using continuous culture with enriched river water, Ellwood *et al.* (1982) reported results opposite to those at first sight expected. It is known that nitrogen-limited (carbon-excess) cultures produce a greater amount of extracellular polysaccharide polymers, which produce thick films on immersed solid surfaces. These films, however, were found to have much lower numbers of bacteria in them than the much thinner films produced under carbon-limited growth conditions. These experiments used a natural mixed culture selected by the culture system itself. The micro-organisms growing attached to the glass surfaces appeared to form micro-colonies of a distinctly different growth form from the free cells growing in suspension. This has been noted in many cases of immobilized cell growth, where SEM studies have been carried out and may well relate to altered metabolism (biochemical differentiation) as well as to the obvious morphological differentiation. In the area of plant cell immobilization, much debate still exists as to whether biochemical differentiation can occur in the absence of morphological differentiation. With regard to immobilized cells, this may be considered as a rather irrelevant discussion since immobilization itself is certain to provoke both types of response. Using a pure culture of a *Pseudomonas* sp. (GAG7) isolated from a surface in the same river, Ellwood *et al.* (1982) were able to study the influence of culture density (by varying the limiting substrate) in both carbon- and nitrogen-limited continuous cultures. SEM studies of bacteria self-immobilized onto glass slides showed considerable numbers of fibrils extending from the bacteria in carbon-limited (glycerol) cultures. In nitrogen-limited (ammonium chloride) cultures fewer, shorter, fibrils were seen but direct cell to cell adhesion was much more obvious. The authors emphasize that the fibrils could only be detected when the more delicate critical-point drying technique was used for SEM specimen preparation. It may be that these processes, which are probably of significance in penetrating electrostatic charge barriers onto clean glass slide surfaces, are generally produced, but not seen due to harsher SEM specimen preparation techniques. The rate of increase of bacterial numbers on the glass surfaces was close to the growth rate of free cells at low culture densities but very much greater at high culture densities. Many of the surface-attached bacteria were present in microcolonies and it did not seem that the results could be explained simply in terms of greater numbers of successful adsorption collisions in the higher cell density cultures.

Powell & Slater (1983) have developed a mathematical model for the deposition of cells onto solid surfaces under laminar flow conditions, extending their previous models of cell desorption under similar flow conditions. The model was based upon results from *Bacillus cereus* grown in a simple chemostat as essentially single cells and then allowed to adsorb onto the surface of glass capillary tubes under controlled conditions. The glass was either in a natural state or treated with dichlorodimethylsilane. Direct microscopic observation with video recording was used to study cell deposition. The residence time on the surface of individual cells could be determined and was shown to be mostly of the order of less than 10 seconds for low flow rates (shear stress of 0.092 N/m^2 but with a significant trend to longer residence times under higher shear stress (0.800 N/m^2). This is explained by the failure of weak adhesive forces to allow any adsorption at all under high shear stresses. Even when weak adhesion does occur it is easily overcome. A maximum population of attached cells was again found and in these experiments the surface coverage by cells onto the untreated glass was only of the order of 1%. After longer periods of time (greater than 43 hours) microcolonies were observed, extending the cell coverage to closer to 5% of the surface. It should be noted that the bacteria were mostly cultured in 1% bactopeptone with 0.5% (w/v) sodium chloride present and attachment studied in this medium. In view of what has been discussed earlier in this chapter, the term 'untreated glass' does not necessarily mean that adsorption of complex organic materials has not occurred during the course of the experiments. The flow pattern over microcolonies was observed to be different from that over a clean glass surface, as might be expected, and this must clearly influence shear rates and presumably cell adsorption close to existing microcolonies. Powell & Slater (1983) go on to develop an equation in which the number of cells on the surface is related, in a dynamic equilibrium, to the rate of cell deposition, the rate of cell removal and the amount of cell growth. Under the conditions they used for the calculations cell growth was negligible on the surface. Their model fitted the experimental data much more closely than an earlier model developed by Bowen, Levine & Epstein. Siliconizing the glass surface led to increased deposition (up to 5% of exposed surface) and an increased rate of deposition. Use of untreated glass surfaces in the presence of an antifoam agent had little effect on the deposition rate but did slightly reduce the number of cells attached. From a study of the critical shear stress (that causing complete removal of all attached cells) the authors speculated that the silicone layer greatly increased the shear stress needed to detach the cells and hence the observed population dynamics were essentially influenced only by cell attachment. This work on *B. cereus* is notable for the elegant simplicity of its direct observational technique, which could be profitably extended to other organisms and situations, and the attempt to provide a quantitative model which might be of use in the design and operation of immobilized cell reactor systems.

Wanner & Gujer (1986) gave a thorough discussion of multispecies models of biofilms and developed an analytical mathematical model of microbial interaction in such films. Various case studies of particular situations were shown to fit the developed model and allow adequate numerical description of the observed phenomena. Ho (1986) discussed the forces involved in the adhesion of micro-organisms to surfaces.

The paper of Ellwood *et al.* (1982) is notable for its exposition of a 'chemiosmotic' approach to the stimulated growth of surface-attached bacteria. This has already been discussed in Chapter 2. One of the interesting findings in their work was a confirmation of the fact that available substrate surface does not appear to be anywhere near saturated by deposition of free cells in suspension growth. Complete coverage appears to occur only by subsequent growth of the originally adsorbed cells, until confluent growth of the microcolonies has occurred. Rutter & Leech (1980) found that *Streptococcus sanguis* showed only 30% coverage of a glass surface at the point when it had reached its maximum saturation with freely suspended cells. Only by subsequent growth was the surface completely covered. Another relevant point is the possibility of synergistic stimulation due to the close proximity in an attached film of mixed species. Several cases are known in which the removal of the extracellular products of one species by another species has this beneficial effect.

Costerton & Cheng (1982) reviewed interactions between micro-organisms growing on surfaces, again drawing attention to the general occurrence of microcolonies and the difficulty of enumerating and studying micro-organisms attached to support materials. They emphasize the serious errors which may arise from the extrapolation of data concerning organisms in free culture to the same species under attached growth conditions. In many natural flowing systems with dilute nutrient supplies (such as streams) the number of bacteria attached to a square centimetre of surface may exceed the numbers found in 500 ml of the water phase. These authors particularly emphasize the experimental aspects of examining surface growth and this has already been referred to in Chapter 2.

A much more pragmatic approach, typical of the vast majority of work on supports for adsorption immobilization, can be represented by the work of Huysman *et al.* (1983) on the selection of porous and non-porous support materials for use in upflow methane reactors. They used a series of model reactors to investigate the colonization of supports with a standard methanogenic sludge inoculum. Gas production was monitored and related back to active biomass concentration in the reactor. Operation was on the basis of continuous recirculation (approximately

45 cycles/day) with one batch feeding every day. The non-porous supports used were sepiolite (a natural clay), zeolite (a natural alumino-silicate), argex (a fire-expanded clay), glass beads and activated carbon. The first four were of approximately 5 mm diameter (but only the glass beads were regularly spherical) and the carbon of around 1 mm diameter. The first three listed are described as having surface pores. The porous materials consisted of natural sponge (50% porosity), reticulated polyurethane foam (of three different pore sizes but each of 97% porosity), unreticulated polyurethane (of about 30% porosity) and a polymethane foam coated with polyvinylchloride. Of the non-porous materials sepiolite was most readily and extensively colonized, which was related to the presence of surface crevices of the size of bacteria. The bundles of needle-shaped crystals can reach 2 μm in length, correlating with the work of Messing previously discussed. Surface area alone, in the absence of appropriately sized irregularities, was not a sufficient guide to colonization, nor was the total cation exchange capacity of the surface. Of the porous materials the reticulated polyurethane foam was most extensively colonized, with the grades T40 and T80 (referring to number of pores per inch and hence relating to average pore sizes of 0.43 and 0.27 mm respectively) were better than T10 (average pores of 2.21 mm). Although the PVC-coated polyurethane was colonized, over the first seven days, at only a slightly lower rate than untreated foam, there was a complete failure of the system after the eighth day. The way in which these practical reactor studies relate to the theoretical discussion of pore size is clear, although the relationship to the details of surface adsorption theory remain typically obscure.

Inspection of the extensive literature on the use of cells immobilized by adsorption onto supports reveals, as expected, a very patchy and unbalanced situation. By far the largest number of researchers have been concerned with the use of very cheap support materials for use in high-volume, low-cost, situations such as wastewater treatment. A much smaller number have been concerned with low-volume–high-value systems, often only at the fringe of possible commercial exploitation, such as pharmacologically active substances from plant cells or alternatives to the conventional processes for antibiotic production. A third, and extensive, area of research deals with growth of animal cell lines. At the research level, much attention has been given to the essentially reverse problem of obtaining good cell growth in the absence of the support surfaces normally obligatory for untransformed animal cells. More recently this has been counterbalanced by the desire to use animal cells to produce potentially useful health care products. This specialized aspect of animal cell immobilization is dealt with separately, in Chapter 6. The rest of the present chapter is devoted to the discussion of a few representative examples, in the context of the principles of adsorption to supports and the methods which have been used to investigate cell adsorption.

Inorganic supports

Micro-organisms which grow on insoluble materials, acting as nutrient or energy sources, are clearly likely to have special properties which make them able to adhere strongly to the appropriate material. DiSpirito, Dugan & Tuovinen (1983) investigated the sorption (as they prefer to call it, avoiding semantic complications by defining it as an all-embracing term without reference to mechanisms of attachment) of *Thiobacillus ferrooxidans* onto particulate material. The supports tested were quartz and fluorapatite (both at 200 mesh size) powdered (flowers) of sulphur, acid-washed glass beads (75–150 μm diameter), pyrite (between 270 and 50 mesh size). For the adsorption experiments washed cells from an early stationary phase culture grown in a medium containing ferrous sulphate were used. Adsorption studies were carried out in 0.055 M sulphuric acid. The cells adsorbed were determined indirectly by estimating protein in the supernatant, after confirming that this gave a reasonably accurate estimate of the biomass adsorbed (allowing also for adsorption onto the glass vessel used). A rapid initial uptake occurred onto glass beads, quartz, fluorapatite and pyrite, reaching a steady state after approximately one hour. The slower rate found with directly added sulphur could be increased to a figure comparable with that of the previous supports by incubating it in spent, resterilized, culture medium. There was an increase in adsorption with increasing amount of support material present, but this did not show direct proportionality and there was clear evidence of complex changes in behaviour, especially with fluorapatite. Likewise, there was not a simple direct relationship between adsorption and cell density. Investigating the influence of pH value (over the physiologically appropriate range of pH 1–4) it was shown that only adsorption to the glass beads was affected, being reduced as the pH value increased. Cells which had been killed (i.e. were no longer capable of growth) using ultraviolet radiation showed slightly lower adsorption rates and the total amount adsorbed was less, particularly for the glass beads and sulphur, but only slightly so for the pyrite. Using dead cells from which the lipopolysaccharides had been removed using 0.5 M sucrose, gave lower adsorption rates and much lower total amounts adsorbed for all three of these particulates. The authors suggest that, except for the glass beads, electrostatic factors were not particularly significant in adsorption, because of the absence of an influence of pH value. It should, however, be remembered that the highest pH value used was 4.0 and this can still be considered as reasonably acid, in the context of the ionization of ionizable groups on the cell surface. Because acid-washed bacteria were used, in the absence of growth substrate, it was concluded that extracellular organic material and growth processes were not directly involved in adsorption. The exact nature of the surface of the bacterium after acid washing is, however, not known and it may certainly be anticipated that there will be extracellular polymeric material present even after acid washing. We are left, as is so often the case in this subject, with a set of interesting results arising from painstaking and lengthy researches, which hint at possible mechanisms of adsorption but are unable to provide definitive conclusions of general applicability.

Myerson & Kline (1983) focus much more closely on the development of a quantitative kinetic model for the same organism, *T. ferrooxidans*, using coal as the support. For those unfamiliar with the field it should be mentioned that a possible use of this organism would be in the bacterial leaching of pyrite sulphur from coal and it is this rather than attachment to coal *per se* which is of interest. In their experiments Myerson & Kline used coal with a 1.66% (dry w/w basis) sulphur content. Protein content was again used to estimate free and attached cells. The coal was used at finer than 250 mesh size with a surface area of approximately 5 m^2/g and experiments were also carried out using controlled pore glass (80 to 120 mesh) of approximately 187 m^2/g as an attachment substrate. Using the coal it was found that at low cell densities all the cells were adsorbed but there was, as expected, a maximum carrying capacity which resulted in free cells remaining in suspension when cell densities were increased. Both these and the controlled pore glass results suggested that irreversible adsorption was occurring. Even modification of the pH value to 10 failed to cause desorption. In all these experiments the bacteria were not acid washed and adsorption was studied in the culture medium (without the iron component). Unfortunately, the pore size of the glass is not specified. The *T. ferrooxidans* is rod-shaped (approximately 1 mm × 0.5 μm). Calculated surface area per bacterium was seven times greater with the porous glass than with the coal and even for the latter less than 0.5% of the surface available was occupied. Their kinetic analysis showed that second order irreversible kinetics provided a satisfactory model of the adsorption process. This is at variance with the conventional Langmuir adsorption isotherm, which considers reversible adsorption processes. Referring back to what was said earlier in the present chapter about the influence of shear stress in the experiments of Powell & Slater (1983), it may well be that cellular adsorption has mixed kinetics of both reversible and irreversible types and that experimental conditions largely determine which one is proved to fit the experimental results. At any rate, once a quantitative relationship has been established it is possible to both predict and test the influence of such factors as cell density and available substrate surface area (of an accessible kind). Such information is clearly relevant for reactor and process design.

Turning now to the vast subject of wastewater treatment, it is difficult to make a selection of what is mentioned specifically. Young & Dahab (1982) studied anaerobic packed-bed reactors. Among desirable characteristics for these is the ability to deal with generally dilute waste streams containing organic matter but to be resilient to sudden higher strength loadings, wide ranges of pH value and possible transient toxic wastes. Low growth rates of the attached biomass and a simple means of removing any excess biomass that does accumulate are also desirable. Self-selecting mixed microbial cultures are generally used, with non-sterile process streams. It can be deduced from this that the actual chemical nature of the support material is likely to be unimportant because it will largely be the flow characteristics of the reactor, the essentially uncontrollable nature of the input process stream and the, likewise, largely uncontrolled nature of the microbial population which will be important. The authors concluded that this was in fact so. The type and size of the support media influence flow and are important in design, but the unit surface area and porosity, as such, did not relate directly to reactor efficiency.

One of the features of interest with the type of reactor studied here is that desirable gas exchange characteristics are related to the production of gaseous products (methane and carbon dioxide) by the attached micro-organisms rather than to the supply of oxygen to the active biomass. Under the conditions described almost any type of support material could be, and no doubt has been, empirically tried, ranging from natural crushed rocks, porous solid wastes, ceramics and other heat-treated inorganics, to natural organic materials and synthetic plastics, in almost every conceivable physical shape and in the whole spectrum of reactor designs. Young and Dahab drew mostly upon results from their own PhD dissertations to compare supports of rock (15–35 mm), Pall rings (16 and 90 mm), perforated spheres (90 mm) and small and large plastic modular media systems. Each material gave an essentially inverse linear relationship between chemical oxygen demand (COD) removal and flow velocity, although the slopes of the lines clearly differed between supports. The spheres, large plastic modular media and 90 mm Pall rings had essentially the same surface areas (around 100 m^2/m^3) and porosities but showed different COD removal characteristics. The authors refer to further results published elsewhere in supporting their conclusions.

It might be concluded that for wastewater treatment it is much more important to get the reactor design and flow characteristics right, with attention to the gross physical features of the support material as this effects the efficiency, rather than be concerned in any detail with the nature of the support material and its finer physical and chemical characteristics or the subtleties of the adsorption process. Such an extreme view, however, is challenged by the results of Messing (1982) who applied his careful observations with controlled pore glass to the problem of producing a high-rate, continuous process for the reduction of organic pollution and the simultaneous production of a high methane content effluent gas. Reporting experiments carried out over two years he describes a two-stage reactor which produced an evolved gas with the remarkable concentration of 90–95% methane in the exit gas. Three different supports were used: extruded Cordierite, extruded brick and insulating firebrick of average pore diameters of 3, 6 and 9 μm respectively. The first two supports were used in rods of approximately 2 mm diameter by 2–6 mm length. The firebrick was cut into slices of approximately 2 mm $\times$ 10 mm $\times$ 20–50 mm. SEM observations showed the bacteria to be immobilized at their ends and to occur in distinct clumps. The results refer to fully operational reactors after active biomass has accumulated. There can be no doubt that selecting materials based on pore size characteristics has resulted in a very efficient process. Because of the novelty of the reactor design, however, some doubt must remain as to whether similar results could have been achieved with other supports used under the same operational conditions. The special features of reactor design are mentioned again in Chapter 7.

Van Den Berg & Kennedy (1981) compared four supports as tubular biomass supports for an anaerobic (methane-generating) fixed-film reactor for wastewater treatment. Two (red draintile clay and grey potters' clay) were inorganic and two were synthetic organic materials (needle-punched polyester and polyvinyl-chloride). Despite the fact that it is frequently used as a packing material for wastewater treatment reactors the polyvinyl-chloride support took considerably longer to reach a maximum pseudo-steady state loading rate, which was also at a considerably lower loading rate than the other materials. The rate of biofilm development was similar for the other three materials but the pseudo-steady state loading values increased in the order grey potters' clay to red draintile clay to needle-punched polyester. The latter is a felt-like material which could trap bacteria. Unfortunately, the critical information is not available on pore sizes and other features which might influence active biomass immobilization levels. The possibility of stimulatory leachates from the red draintile clay (and to a lesser extent the potters' clay) is, however, mentioned by the authors. COD conversion efficiencies were similar with all four supports but methane production per metre of gross support surface increased in the order polyvinylchloride to grey potters' clay to needle-punched polyester to red draintile clay. It is, however, certainly obvious that this area of research could benefit from some more fundamental studies. The relatively low profits and massive investment costs for large-scale wastewater treatment plants appear to have encouraged too great a concern for mere solid, fluid and gas handling solutions and too little a concern for the biological aspects of the cell immobilization, which is the functional heart of the process.

Turning now to processes in which the biological aspects are more central, Marcipar *et al.* (1979) considered the immobilization of yeast onto clay-based ceramic supports. They used *S. cerevisiae, C. tropicalis, Rhodotorula* sp. and *Trichosporon* sp. Cells were recovered from liquid culture on 3% malt extract by centrifugation and resuspended in phosphate buffer. The support used was Biodamine^R (particle size 1.25–2.5 mm and 7–11% apparent porosity) packed into a column with recirculation capabilities. The percentage of cells adsorbed from a 2×10^8 cells/ml suspension was found to be characteristic of the organism. Values of around 75% were found for *S. cerevisiae* and *C. tropicalis*, with only a few per cent less when at pH 6 instead of pH 4. For *Trichosporon* sp. the value was 50% at pH 4 and 47% at pH 6, while for *Rhodotorula* sp. it was 65% and 39% respectively. Minimum flow rates causing detachment for *S. cerevisiae* and *C. tropicalis*, were 4.7 cm/min at pH 4 and 3.4 and 4.7 cm/min at pH 6 respectively. There was evidence, using a Coulter Counter, that larger, older cells attach preferentially. Oxygen uptake rates were increased after immobilization but no values were obtained for ethanol production.

Bland *et al.* (1982) used fine particles (250–380 μm size) of vermiculite to immobilize *Zymomonas mobilis* for the high-rate production of ethanol in their AFEB (attached film expanded bed) reactor, of inverted cone configuration. After operational cell loadings had been reached the vermiculite particles were shown to be nearly covered with bacteria, with cells between the vermiculite layers. Flocs (approximately 1 mm diameter) were also present in the reactor. Maximum productivity occurred at a dilution rate of 3.6 with 64% conversion of the substrate. Further optimization could probably be achieved using other operational conditions and support materials based on pore size selection principles.

Arcuri *et al.* (1980) compared ethanol production by *Z. mobilis* immobilized in the presence of polystyrene beads (1 mm diameter) and into glass fibre pads (approximately 2 mm $\times$ 2 mm). Bacteria were allowed to self-adsorb into the glass fibre and growth was evident in the pads. Negligible attachment was found to the polystyrene beads but the bacteria readily formed flocs in the presence of the beads. The self-flocculating system appeared to have a greater productivity (in terms of reactor volume), but true comparison was not possible because of differences in reactor design between the two systems and absence of data on active biomass concentrations. Two points of particular interest in the present context arise from this research. One is the ease of immobilization into fibre-based pads (which could be prepared in more satisfactory shapes), supporting the suggested suitability of such a system as proposed by Atkinson *et al.* (1980). The other is the fact that potential support materials may influence cell properties (in this case self-flocculation ability) without themselves necessarily acting as direct immobilization supports. The mechanism for this is unclear.

An alternative approach to the physical shape of immobilization supports is exemplified by the work of Ghommidh, Navarro & Durand (1981) on acetic acid production by *Acetobacter aceti*. They used the porous ceramic Cordierite but in the form of large cylinders (described as monoliths in the paper) of 5 cm × 10 cm, with square section (1.15 mm × 1.15 mm) longitudinal holes passing through them, held in a down-flow column reactor. The bacteria self-immobilized easily and permanently, following first order kinetics until saturation was reached. Used for continuous acetic acid production they reported very favourable productivities. Due to the rather small size of the channels this particular system would probably be only suitable for organisms which do not over-produce fresh biomass growth during operation, but the possible convenience of a 'replacement canister' type of immobilized reactor certainly deserves consideration. In a later paper Ghommidh, Navarro & Messing (1982) go on to explain in detail the situation for this bacterial system when acetic acid exerts a growth inhibitory effect, emphasizing the value of immobilized cell systems compared with conventional continuous culture systems, for the production of metabolites causing growth inhibition. At this point it is worth mentioning the potential value of such immobilized cell systems for the removal of toxic substances from wastewaters. Suidan, Cross & Fong (1980) for example describe a self-regenerating continuous anaerobic reactor for the degradation of catechol using a support of granular activated carbon. The inherent adsorption capacity of the carbon for organic substances appears to co-operate in a favourable manner with the active biomass, allowing longer contact times for biological degradation to occur. There is no doubt that greater use could be made of carbon as a support. There is virtually no limit in the possibilities of novel support systems.

Following this theme Jian *et al.* (1983) devised a means of making porous Kaolinite beads by mixing with rice hull ash and heating in a furnace at 800 °C for five hours. With the correct proportions very porous beads, with good mechanical properties, could be made. As these authors, and others before, point out, the choice of support is largely a matter of guesswork. Whether this situation is improved by a new support medium, which involves collection of rice hulls from the local commune and ashing by ignition of a heap one metre in diameter, remains to be seen. It does illustrate, however, the immobilization of micro-organisms. In this particular instance methane-generating organisms are reported to have been successfully immobilized. Hancher & Perona (1982) used particles of coal (30–60 mesh) to support denitrification bacteria in a double unit fluidized bed reactor for the removal of nitrate from wastewaters. The system operated continuously for seven months and a kinetic model based on Michaelis–Menten relationships was developed.

Whether coal should be described as a mineral or an organic support is left to the reader but to conclude this section on inorganic supports the work of Gbewonyo & Wang (1983) picks up again the beneficial process characteristics gained from immobilization. They immobilized spores of *Penicillium chrysogenum* into carefully selected and sized beads of celite. Great care was taken (a model for other researchers) to obtain thorough details of the support material, which had been selected on the basis of extensive studies. Celite (fossilized shells of diatoms) has a greater than 70% void space and this was considered important for the present application. The pores in the size range 1–15 μm (diameter) were considered most important, with the spores being of 2 μm diameter and the hyphae of 3.5 μm diameter. High hyphal biomass densities could be produced by growth inside the beads, to yield up to 1.8 times the free culture biomass when 10% (w/v) celite beads were used. This was attributed to easier oxygen transfer because the increased viscosity of free mycelial cultures was not found with growth inside the celite beads. Since the method is described as being independent of strain and media conditions the authors suggest it will have many applications. Arcuri, Slaff & Greasham (1986) described the successful continuous production of the antibiotic thienamycin by *Streptomyces cattleya* self-immobilized into celite. This simple micro-scale version of the Atkinson *et al.* (1980) concepts will undoubtedly be the subject of much future research.

Organic supports

One of the original materials for the immobilization of micro-organisms is the natural, complex, organic material known as wood. Thin shavings of wood were traditionally used as support in the production of acetic acid by *Acetobacter*. At first sight wood is an attractive material to use. It is cheap and available almost anywhere and can be reduced to coarse or fine particle size with commonly available equipment. It is manifestly porous and is readily wetted and absorbs water. Microscopic examination shows many of the pores to be of an appropriate dimension for microbial cell immobilization. Certain major problems do exist, however, in using wood as an immobilization support. As anyone with an elementary knowledge of botany knows, wood is not a homogeneous material. It is, in fact, a tissue (system) made up of several different cell types. A major difference in structure exists between the softwoods and the hardwoods. These terms do not relate in any way to mechanical properties but to the fact that the former come from gymnosperms and the latter from angiosperms. In the hardwoods, but not the softwoods, fine holes (pores

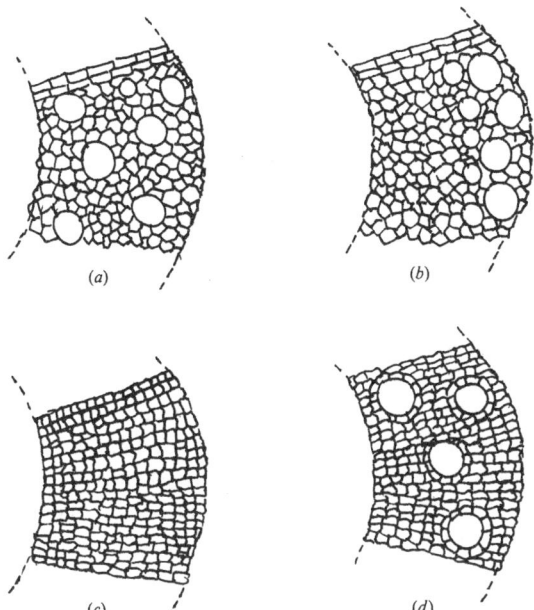

Figure 5.2. Diagrammatic representations of some of the important differences between wood from various types of tree species. Only a portion of a single annual ring is shown in each case. (*a*) Diffuse porous hardwood; (*b*) ring porous hardwood; (*c*) ordinary softwood; (*d*) softwood with resin ducts. In all cases the amount of accessible pore space will depend on the particle size.

as they are known and appear in transverse cross-sections of wood) pass completely through the wood in a longitudinal (with respect to the original plant stem axis) direction. These holes are the cavities inside the interconnected cells forming the vessel system, used originally in the living plant for the conduction of water. During development the end walls of the cells are lost leaving an open tube system. Softwoods do not have vessels but instead have only the elongated cells (with complete end walls) known as tracheids. Tracheids are also found in hardwoods. The internal cavity of tracheids is opened, in a gross way, to the outside, only when the wood is cut and the cells damaged (as is true, of course, for vessels as well) but the cavity does not extend completely through the particle unless this is less than the length of the tracheid cell itself. The diameter of the cavity within vessels varies enormously between species, as does the way in which the vessels are distributed amongst the other cell types making up the xylem tissue. At the level of one annual ring (seasonal growth cycle) the differences may be very marked in 'ring-porous' woods and not visible in 'diffuse-porous' woods (Figure 5.2).

The majority of woods of commercial significance can be identified, often to the species level, by the microscopic examination of the types and arrangement of cells within the wood. Another species-specific feature and one which may, in fact, vary within a single species and even different parts of the same tree, is the exact chemical composition and physical properties of the cell walls. The major components of tracheid and vessel cell walls are polysaccharides, such as cellulose and the hemicelluloses and the complex polyphenolic material known as lignin. The exact chemical natures of the hemicelluloses and lignins differ between species. The relative proportions of the three major fractions of the wall differ between and within species. Gerson & Zajic (1979) draw attention to the variation in surface free energies between different natural woods (generally in the range 40–58 ergs/cm^2). Another significant feature of wood, which again varies from species to species, is the presence of a whole range of soluble or partially soluble ('extractives' as they are known) substances with toxicity towards micro-organisms. It is this which, in major part and taken with the cell wall composition, will determine the durability of the wood – its resistance to microbial degradation. At the same time it will possibly cause serious problems during the use of wood as an immobilization support for micro-organisms used in industrial processes.

By careful selection and specification of the source of a natural wood-based support it would be possible to overcome these problems. These problems are unfortunately not always clearly explained. Michaux *et al.* (1982), for example, consider it important to specify the source of the various soluble polymeric substances used in their paper (discussed earlier in this chapter) dealing with cell immobilization in relation to altered zeta potentials. They do not, however, give any useful information about the 'sawdust' used as the support, referring only to an earlier paper. In practice, it often proves remarkably difficult to obtain a reliable supply of sawdust from a specified tree species. The researcher may have to produce his own supply, which on a large scale often proves unexpectedly laborious. In uncontrolled mixed-culture studies the use of wood may lead to its colonization and degradation by one or more of the many microbial species capable of using it as a carbon source. Even species which are not cellulolytic or lignolytic, in the usual meaning of these terms, can obtain some growth substrate from a material as chemically complex and diverse as wood. It has been suggested that better adhesion occurs (to plastic) if microbes are able to obtain part of their nutritional requirements from the support (Hollo *et al.*, 1979). This may also be true for wood-based supports.

The amount of space devoted here to wood is deliberately out of proportion to its actual use as an immobilization support. This is because it exemplifies, in an extreme form, the difficulties which can arise in the use of natural organic supports. These are the major ones of consistency of properties, reliability of supply and product specifications, and possible biodegradation in use. These, of course, are in addition to the question of whether they are actually efficient supports in comparison to alternative materials. Many natural materials can be obtained with more consistent properties. Coarse natural 'hard' fibres, already used commercially, often consist almost entirely of one cell type (elongated, heavily lignified, thick-walled fibre cells). These are generally very resistant to microbial degradation and their preparation often involves a stage in which other cell types are microbially or enzymically removed. Local materials, for example from the primary stages of food processing, may well show useful characteristics for 'low technology' processes. Cotton fibres are almost pure cellulose and there is a reliable supply of cellulose in all manner of physical forms. There is an enormous literature on the degradation of cellulose by micro-organisms, covering also the attachment of cellulolytic microbes to the insoluble cellulose. It would not be appropriate to discuss this in the present book although it is clearly of some peripheral interest to the use of supports for cell immobilization. Suffice it to say that there has not been a very extensive use of particulate celluloses for immobilization but there has been much more work on various derivatives prepared from cellulose. These may be regenerated into membrane form, for use in membrane and hollow-fibre reactors, used as the basis of ion-exchange materials or even used as supports in their own right.

Chen & Zall (1982) used cellulose acetate as support in a continuous attached film expanded-bed reactor for the fermentation of whey to alcohol using *Saccharomyces fragi*. They record that several other (unspecified) materials were investigated and that previous observations with this species of yeast, in a reverse osmosis plant, had shown its natural immobilization to cellulose acetate membranes. Unfortunately, the details given in the paper leave much to be desired. The fermenter was filled with growth medium and approximately one third of its volume with cellulose acetate fibres. The highest productivity was obtained at a dilution rate of 1.1 which was almost four times the maximum specific growth rate in batch studies of the same medium – a clear practical indication of successful immobilization. The cellulose acetate was not in the optimum physical form because it showed poor settling characteristics and evolved carbon dioxide bubbles tended to carry support and attached cells out of the fermenter. The authors also carried out the same fermentation in a membrane fermenter and suggested that both systems might have commercial potential. Sometimes it may be difficult to decide whether the self-immobilization of cells is an advantage or a disadvantage. De Cabrera *et al.* (1981), for example, investigated the direct extraction and conversion of fermentable substrates from sugar cane segments by yeast. The EX-FERM process involves replaceable packed-bed columns of the sugar cane and a circulation system from a fermenter/reservoir. It was found that as much as half (and probably more) of the yeast became trapped in the sugar cane segments. Because the columns are changed when fermentable sugars are exhausted it means that a considerable quantity of yeast biomass is continually being wasted from the system. This could be considered a serious loss of active biomass or a beneficial nutritional upgrading of the spent sugar cane residue, depending on the viewpoint. Many other processes involving adsorption of cells to natural solid organic materials could be discussed but they would shed little light on the principles or practice of immobilization. As in the case just mentioned, they are essentially fortuitous results arising from empirical studies carried out for other purposes.

For the reasons which have been discussed earlier in this chapter the vast majority of those who deliberately attempt adsorption immobilization prefer to use reproducible synthetic organic materials rather than natural materials. The possible range of synthetic materials is clearly enormous. Some have been particularly fabricated as biomass support surfaces and a few have already been mentioned in the discussion of comparative studies which also involve inorganic support materials. Genung, Hancher, Rivera & Harris (1982) studied the ANFLOW system of municipal wastewater treatment being developed at Knoxville, USA. This involves an anaerobic up-flow, packed-bed reactor. Among the several objectives of the reported work was the evaluation of a plastic support material: 3 inch diameter Pall rings made of polypropylene. It was chosen for its cheapness, the high reactor void volume obtained when randomly packed in the chosen physical form, its relatively high surface to volume ratio and open structure (in comparison to earlier used Raschig rings). It was also considered important that it was commercially available and from a supplier already active in the appropriate industrial sector. These latter points are often neglected by those carrying out small laboratory trials. In the present case the reactor is of substantial size (4.9 m × 5.4 m). Standard anaerobic sludge was used as the inoculum. Without considering the fine details of their results, which are not of particular relevance in the present context, it can be stated that the reactor performed well over many months, with a residence time (which could certainly be reduced) of approximately one day, using a weak municipal wastewater. Excessive sludge build-up, requiring solids removal, did not occur over a nine month period of operation. Obviously a variety of supports might prove quite suitable in this type of application. In earlier work on the process a ceramic support had been used but was considered too expensive for use in competition against alternative processes. The authors promised comparative cost, energy requirement and engineering analyses at a future date.

Kennedy & Van Den Berg (1982) continued their work on the needle-punched polyester fibre support mentioned earlier in the present chapter. In the case now under discussion the effluent was piggery waste. The particular merits of their downflow stationary fixed-film reactor system are described as the ability to cope, without reactor overload failure, with short-term pulses of very high organic loading. This resilience of an immobilized system is particularly valuable where a highly variable effluent stream supplies the reactor. This would typically occur during a daily piggery cleaning regime. The support material was used as an orientated film (375 g/m^2) in vertical channels of approximately 2.8 cm cross-sectional dimension. A continuous (actually 48 batch feeds/day) operation regime was compared with a 1 batch/day feed of the same total quantity. The single batch loading resulted in a comparable cyclical methane production peak a few hours later. Since the increase was as much as 50% from the daily average methane production rate this could provide a single means of matching energy supply to demand, which is always a problem with biogas generation systems. After 14 months of operation solids accumulation had not been a problem and channel blockage had not occurred. The support material had not deteriorated and was evenly colonized (to a thickness of a few mm) by the active biomass. The same support has also been shown to be satisfactory at thermophilic (55 °C) temperatures for wastewater treatment applications.

In a novel application of *Thiobacillus ferrooxidans* for the oxidation of iron from the ferrous to the ferric state Livesey-Goldblatt, Tunley & Nagy (1977) compared four support materials. Firstly, upright perspex sheets were sprayed with a ferrous salt solution, relying on natural air diffusion through the liquid film on the sheets. Secondly, a horizontally rotating polyurethane disc system was immersed to 35% of disc surface area in the solution. Thirdly, a honeycomb of ribbed corrugated PVC (generally available for sewage treatment plant use) was completely submerged in the solution which was kept fully aerated by a suitable aerator. Fourthly, PVC rings were randomly packed into a column and the ferrous salt solution sprayed over the top, relying on natural air diffusion through the film. Complete active films were built up on the surfaces before the reactors were started. The highest oxidation rate was found with the corrugated PVC system, followed by the perspex sheet/spray system and then the polyurethane rotating disc reactor. The PVC ring trickle flow column gave the lowest oxidation rate. Bacterial films were stable on all supports and the better performance of the corrugated PVC system was clearly attributed to the much higher oxygen levels obtained with the forced aeration system. Once again the conclusion must be drawn that it is the reactor design and operational regime that inevitably emerge as the factors of most importance. Only after this has been fixed can comparisons between support materials be carried out and, of course, a support appropriate for one design of reactor may be unsuitable for another.

Hollo *et al.* (1979) considered the use of immobilized micro-organisms for the denitrification and removal of heavy metals from wastewaters. They draw attention to the high hydraulic resistance of finely granulated inorganic carriers and the generally accepted solution of using plastic supports instead, in fixed-bed reactors. In their experiments, Hollo *et al.* used *Pseudomonas aeruginosa* and sought ways of ensuring a firm attachment of active biomass to the support. After comparing a wide range of potential plastic support materials they chose two. Polyvinylchloride (PVC) film contains added 'softeners' which can act as a carbon source for microorganisms. In the process of assimilating such 'softeners' the bacteria achieve a very firm surface attachment, even to the extent of burrowing into the film itself. The PVC sheet was cut into small squares and used in a packed-bed reactor. The plastic itself could serve as a sole carbon source but film development was improved by supplying an additional soluble carbon source and nutrient supplements. Surface conditioning of the plastic by these was also suggested as a mechanism stimulating attachment. Good surface colonization was observed with the SEM. The second support selected was melt-blown polypropylene in the form of fine filaments (under 1 μm in diameter) which has a high surface to weight ratio. The suitability of the surface for bacterial attachment was improved by a few seconds' treatment in an oxygen plasma generated in a flow discharge apparatus. This greatly improved surface wettability and surface charge. A minimal release of bacteria was found but support columns were run in series, with the older being removed when the new one was fully saturated with active biomass. It was concluded that the PVC support was adequate for denitrification, giving a higher viability but a rather poorer film stability than the polypropylene. For the removal of plutonium from wastewater streams the fibrous polypropylene support was more satisfactory. From the results given it is impossible, however, to decide whether the physical form or the type of plastic is the aspect of major significance. The various plastics have different physical and mechanical properties which make them suitable for certain types of fabrication process.

Blain *et al.* (1979) used high density polypropylene with a roughened surface for the discs of their improved horizontal rotating disc fermenter. Good growth was obtained with a wide variety of filamentous fungi. Under conditions of high substrate supply *Aspergillus niger*, for example, produces firmly attached mycelial layers of 5–10 mm thickness. Despite this depth, and an observable heterogeneity through the layer (as shown by the C:N ratios and oxygen uptake rates), the innermost parts remained metabolically active. The mycelial film remained firmly attached for long periods of operation. Spore formation did not occur unless areas of the mycelium were allowed to become dry. The potential use of the disc system for citric acid production by *A. niger* was studied by the same authors in a later paper (Anderson *et al.*, 1980).

A wide variety of irregularly-open, fibrous, mesh systems are available and many of these have been tested for use as supports. Initial cell uptake is by adsorption and subsequent growth into the interstices may result in a proportion of self-entrapment into the support. Rhodes, Robins & Payne (1982) reported on their experiments with plant cells grown in suspension culture. When mixed with pieces of sterilized nylon mesh pads (sold originally as pan-scrubbers) the cells were rapidly adsorbed into the mesh and grew within the mesh at the same rate as free cells. High cell loadings could be obtained. Lindsey & Yeoman (1983) combined this with entrapment inside the nylon mesh in agar or alginate gels, providing a gentle system but with better mechanical stability than the entrapment gels used alone. The immobilized cells appeared to reach a stationary growth phase more quickly than in free suspension culture. This may be of great benefit for the production of commercially interesting secondary products. The authors conclude that the benefits of immobilization may arise from close association of the plant cells, under conditions whereby the cells do not move relative to one another. Such an arrangement cannot be obtained by suspension culture, except in the uncontrollable formation, during growth, of irregular clumps of cells, which, for various reasons, is considered to be a serious problem rather than a benefit. Immobilized plant cells are discussed by Morris *et al.* (1986).

Fibrous nylon net is a convenient support for cell self-immobilization. The technique is simple and mild and the support cheap and readily available. Linko *et al.* (1986), for example, used it for the immobilization of *Pichia stipitis* during studies of ethanol production from D-xylose.

The use of porous foam plastics has already been mentioned in the section on inorganic supports as forming part of several comparative studies on immobilization. They have been particularly studied as supports for mixed cultures used in wastewater treatment plants. An interesting plastic foam-based system has been patented, deriving from the principles expounded by Atkinson *et al.* (1980). This involves the periodic removal of the freely suspended foam support particles and the extrusion from them by squeezing between rollers, of excess biomass. Because this is in a highly concentrated form, high in solids, it is easier to handle and dispose of than excess free-floating biomass carried out in the outflow of a treatment plant. In their studies of upflow methane reactors Huysman *et al.* (1983) found high porosity, reticulated polyurethane foam a good support material. The type of foam has good mechanical stability and is used, for example, in some air-filter systems. It is therefore available in sheet form in a range of thicknesses and pore sizes, to reproducible specifications. For these reasons alone it is likely to find increasing use as a biomass support system.

In an interesting application of foam immobilization Muallem, Bruce

& Hall (1983) studied the photoproduction of hydrogen gas and reduced NADP by cyanobacteria. In flask experiments three types of polyurethane foams were used, cut into small pieces (approximately 5 mm × 5 mm × 5 mm cubes). Two polyester types (from different suppliers) and one polyvinyl type were tested. The species studied were *Chlorogloea fritschii, Nostoc muscorum* and *Mastigocladus laminosus.* Foam-grown cells of the last species were more active and showed better long-term stability of hydrogen production than free-grown cells, especially when one of the polyester foams was used. Cells of *C. fritschii* required permeabilization, which was done by freezing and thawing the cells, while immobilized in the foam. The authors consider the polyurethane foams to be preferable to alginate immobilization and to retain biomass well, without washout, in continuous flow systems. Improvements in foam specifications were thought possible. The use of foam immobilization for light-requiring cells is receiving attention from several researchers and promises to open up some exciting new developments.

Work with these simple methods of immobilization for plant cells continues in many laboratories. Mavituna & Park (1985), for example, used porous polyurethane foam to immobilize *Capsicum frutescens.* Growth patterns and final cell yields were similar to those of free cells.

Ion-exchange supports

Because ionic charge interactions are clearly important in the early stages of adsorption it is an obvious step forward to use ion-exchange materials as supports for cell immobilization. Because they are commonly available in a wide range of types, for both laboratory and industrial uses, such experiments are easily set up. The major disadvantage of the use of ion-exchange supports is their high cost compared to many other adsorption materials. Many natural organic and some inorganic materials could be considered to be ion-exchange materials, because they do have ionizable groups on the surface and may even exhibit high cation exchange capacities. The supports considered in the present section, however, are those available from normal chemical suppliers specifically as ion-exchange materials. The ion-exchange resins, as the name suggests, are based upon organic polymers. A typical matrix is polystyrene cross-linked with divinylbenzene. The water regain of the dry resin beads causes the beads to swell in size when immersed in an aqueous solution. The degree of cross-linking of the matrix has a major influence on the water regain and the ion-exchange resins are available in a wide range of types, falling into two major classes. The gel type has self-obvious microporous structures while the macroreticular types have large interconnected pores, within the normally spherical resin beads.

Table 5.2. *A simple classification of ion-exchange materials*

Support types	Typical Functional Groups[a]	
	Acidic	Basic
Resins		
Polystyrene		
Polyacrylic	—SO$_3$H	—N·(CH$_3$)$_3$·OH
Polymethacrylic	—COOH[b]	—NH$_2$[b]
Macroreticular		
Polysaccharides		
Celluloses		
CM	—CH$_2$·COOH	
P	—PO$_3$H$_2$	
DEAE		—CH$_2$·CH$_2$·N·(C$_2$H$_5$)$_2$
Sephadex		
CM	—CH$_2$·COOH	
SM	—SO$_3$H	
DEAE		—CH$_2$·CH$_2$·N.(C$_2$H$_5$)$_2$

[a] Exchange capacity depends on the degree of substitution.
[b] Weak functional group means narrower pH range of use.

The average pore size can be controlled. Earlier types of ion-exchange resin did not have the macroreticular structure now available and it used to be considered that cellulose-based ion-exchanges should be used for large molecules, and by implication this would be extended to cell immobilization. This stricture no longer necessarily applies.

Considering now the available ionizable groups which may be introduced onto the resin matrix leads firstly to a classification as anionic or cationic, depending on their affinity for either negatively- or positively-charged ions. A further subdivision can then be made in each group into strongly and weakly ionized exchange groups. The strongly ionized groups will exist in this state independent of the pH (except at very extreme pH values). The weakly ionized groups are effective only over relatively narrow pH ranges, where exchange capacity is at a maximum due to fuller ionization. The presence of salts (ionic strength) of a solution will influence ionization. The concentration of ionizable groups in a resin may correspond to as much as 10 M strength and hence may exhibit a very high exchange capacity. Since the surface charge of almost all micro-organisms is negative it follows that such cells will attach more readily to the surface of positively-charged (anionic) ion-exchangers (Table 5.2).

The ion-exchange celluloses carry ionizable groups as substituents on the hydroxyl groups of the glucose subunits. If too many hydroxyl groups are substituted the derivatized cellulose becomes water-soluble and so it is only possible to have a relatively low exchange capacity on insoluble celluloses. The substituent groups are also generally rather weakly ionized. A further point which may restrict the use of the ion-exchange celluloses in some application is their potentially biodegradable matrix and more fibrous physical form, which is not as universally desirable as the spherical bead form usually used for ion-exchange resins. It should, perhaps, be noted that the various ion-exchange materials have been developed for chemical and biochemical applications, especially for deionization and for use in chromatographic separation, and not specifically for cell immobilization.

Kolot (1981) discusses both types of ion-exchange material. Because microbial cells can be made to exhibit a dipolar nature, depending on the pH value of the solution, it is possible for them to bind to both cationic and anionic materials. In practice, however, cell adsorption is generally very poor on most types and only a few are generally recommended. These are the strongly basic anionic exchangers. It should be noted that the individual suppliers have adopted particular trade names and numbering systems for their products and this makes it essential to consult trade literature carefully, particularly when seeking similar materials from different suppliers. Microbial cells are rapidly adsorbed, with little leakage and under mild conditions, to ion-exchange materials. Factors influencing attachment include culture age, cell pre-treatment, medium composition and pH value. After immobilization various physiological and metabolic changes may occur. Differences in the morphology of the cells, particularly any new ones produced by subsequent growth may also occur.

The production of ethanol by *S. cerevisiae* was investigated in a frequently quoted paper of Daugulis *et al*. (1981). They compared adsorption onto three gel-typed anionic resins (two strongly basic), five macroreticular type anionic resins (all strongly basic) and the two simpler supports, activated carbon and crushed ceramic berl saddles. One type of macroreticular ion-exchange resin was clearly superior as an adsorbant with well over 100 mg cell dry wt/g dry support. A second type of the same category gave almost 50 mg/g. The best of the gel-type exchangers was the weakly basic one, with just over 10 mg/g. All the rest of the resins and the two non-resin supports gave very poor or negligible adsorption. In further experiments with the bacterium *Z. mobilis*, Krug & Daugulis (1983) again used eight ion-exchange resins (not all the same as those used for the yeast work), activated charcoal and ceramic (described as crushed brick). One resin was strongly acidic (cationic) and one weakly acidic (cationic) and of the six basic (anionic) resins five were strongly basic. The two best materials for adsorption were the same as for the yeast but in the opposite order and much lower (approximately 50 and 37 mg dry wt cells/g dry support). A clear age effect of the culture was seen, with younger cell cultures being much better adsorbed than older ones. Using the SEM it was seen that the form of the bacterium changed from single short cells to multiple and filamentous forms during a reactor run. After a 200 hour run the column had developed sufficient excess biomass to hamper severely the flow in the column reactor being used. It was suggested that nutrient limitation might overcome this problem.

The use of various ion-exchange celluloses is discussed by Kolot (1981), reporting the work of others. For spores of *Aspergillus oryzae, Penicillium roqueforti* and *A. wentii* the best support was ECTEOLA-cellulose (an anionic exchanger with mixed amines as functional groups), at both pH 5 and pH 9. The other ion-exchangers used were DEAE-cellulose (anionic with diethylaminoethyl groups), P-cellulose (cationic with phosphate groups) and CM-cellulose (cationic with carboxymethyl groups). *A. oryzae* spores were reasonably well adsorbed to all of these, almost regardless of pH value. Spores of the other two species were very variably adsorbed, depending on the pH value and which of the three other ion-exchangers was being used. As with the ion-exchange resins, different manufacturers may use their own trade names and codes for essentially similar materials and many types of ion-exchange celluloses are available. The trade literature must be consulted but is likely to be relatively uninformative with regard to their use for cell immobilization.

6   Special methods

One of the major problems encountered in the use of immobilized cells as biocatalysts is the hindered diffusion of substrate and products. Part of this arises from the technique of immobilization itself. It can be minimized by appropriate selection of the immobilization technique and the physical shape and size of the biocatalyst preparation. Comparative studies of free and cell-bound (or enclosed) enzyme activities, however, consistently show an increased value for the Michaelis constant in the latter situation. This is generally considered to be due to diffusion restriction, reducing the substrate concentration and increasing the product concentration at the active site of the enzyme molecules, rather than being due to a change in enzyme conformation. A further problem which arises with many cells is the failure of the cell to excrete the desired product to the outside, even where the substrate, or precursors for more complex biosynthetic processes, can readily enter the cells.

Several different physically distinct diffusion barriers can be recognized. With plant and microbial cells the cell wall clearly hinders free diffusion of larger molecules, although its effect may be relatively insignificant with smaller molecules. Chapter 2 should be consulted for more information on this point. The preparation of protoplasts by enzymic dissolution of the cell walls is a well-established technique in cell biology. It cannot be considered a permanent solution because viable cells will regrow their walls, but it does allow certain benefits to be obtained, as discussed later in this chapter. Animal cells, of course, do not possess cell walls and require special handling during immobilization. For reasons discussed in Chapter 1, some animal cells may normally grow only in an immobilized state and special cell carriers have been developed for their bulk culture.

Any cell considered to be viable must possess a selectively permeable plasma membrane. The definition of the term 'viable' has been explored in Chapter 1, together with a consideration of the selective permeability properties of cell membranes. During the preparation of immobilized cells the properties of the plasma membrane may be altered, in the direction of increased permeability. This may be merely the consequence of choosing a particular technique of cell immobilization, or from part of a known procedure leading to a non-viable immobilized cell preparation. Such non-viable preparations are of major economic significance and are reviewed, in relation to single and 'few-step' enzyme reactions, by Gestrelius (1983). The term 'permeabilization' is used to describe a deliberate treatment to increase membrane permeability. The harsher treatments result in complete disruption of the membrane and total loss of selective permeability. The gentler methods may retain structural integrity (as seen by transmission electron microscopy) but have demonstrably altered permeability properties.

Cell disruption under mild conditions is used to prepare subcellular organelles with specified metabolic activities. The techniques are well described in conventional biochemical texts. Such organelle fractions can be immobilized to yield preparations enriched in particular biosynthetic pathways, with competing or undesirable side reactions removed during the preparation. The diffusion barriers of cell walls and plasma membranes are automatically removed during organelle preparation and the organelles, unlike viable protoplasts, are incapable of regenerating a cell wall. Immobilization may considerably stabilize an organelle preparation.

It sometimes happens that an immobilized cell preparation does not possess the full range of biocatalytic activity required for a desired bioconversion process, due to the absence of the appropriate structural genes in the cells. This could clearly be solved by selection of a mutant strain or recombinant DNA techniques. Both of these solutions require time and appropriate technical expertise in genetic manipulation. An alternative and cheap, if perhaps empirical, approach is to co-immobilize cells and exogenous enzymes together. This approach to biocatalytic inadequacy is a fertile ground for experimentation.

The remainder of this chapter is devoted to a brief individual overview of the topics outlined above.

Permeabilization

In its severest form the process of permeabilization is sometimes called 'cell rupture' (Gestrelius, 1983). Three distinct time periods for permeabilization exist: before, during or after immobilization. In all cases a variety of treatments may be used (Table 6.1).

Prior to immobilization animal cells, protoplasts and membrane-bounded organelles can be subjected to osmotic shock (i.e. placed in a solution of lower osmotic strength than the contents) and allowed to burst. This is an easy and mild technique, retaining most of the enzymes in an active state, but allowing the soluble enzymes to leak way into the supernatant phase during recovery of the cellular debris, which is the fraction to be immobilized. Conventional cell disruption, by mechanical homogenization, sonication or high/low pressure cycling through a narrow orifice can all be successful, depending on the type of cell being

Table 6.1. *Some methods of cell permeabilization*

Technique	Advantages	Limitation
Mechanical disruption	No toxic chemicals used Familiar technique	Loss of viability Leakage of enzymes Inactivation of enzymes May be difficult
Heat drying	No toxic chemicals used Cheap Quick Easy	Loss of viability Inactivation of enzymes
Freeze drying	No toxic chemicals used Mild Easy Viability may be retained	Special equipment needed Rather slow process
Freezing	No toxic chemicals used Cheap Quick Easy	Loss of viability Leakage of enzymes on thawing
Harsh immobilization techniques	No additional treatment needed	Loss of viability Inactivation of enzymes
Organic liquids (solvents)	Cheap Quick Easy	Loss of viability Inactivation of enzymes Possible carry-over to products
Surface active agents	Cheap Quick Easy	Loss of viability Leakage of enzymes Inactivation of enzymes Possible carry-over to products

used. The possible problems of enzyme inactivation due to uncontrollable temperature rises, and the difficulty of treating large quantities of cells, are well-known. Some cell walls are exceptionally tough and difficult to break open. In appropriate cases, cell walls may be softened and loosened by treatment with hydrolytic, cell wall-degrading enzymes. A treatment regime and/or conditions which fall short of those required for protoplast preparation, can result in a relatively mild, osmotically disrupted preparation, retaining higher enzyme activities than those obtained by the more conventional cell disruption techniques mentioned above. Enzyme treatments which cause partial disruption of the cell wall will increase its reactivity to immobilization reagents by exposing greater numbers of reactive groups. At the same time they will generally increase the sensitivity of the membrane-bounded protoplasts to damage by the immobilization reagents and this needs careful consideration in designing the experimental protocol.

The term 'drying' covers a wide range of techniques which can result in permeabilization. On a small scale, simple air-drying of harvested cells can be used but could present a dust hazard when using micro-organisms. Where drying results in loss of viability, then permeabilization is certain to have occurred. Freeze-drying conditions can be adjusted to give negligible cell damage, with retention of cell viability and absence of permeabilization. Under conditions not optimized for retention of viability it is possible to achieve permeabilization. Because the conditions during freeze-drying can be accurately controlled there is ample scope for the selection of an appropriate protocol for permeabilization. The major problem, in industrial terms, is probably the cost of freeze-drying. A wide variety of heat-assisted drying equipment is available for both batch and continuous use. In general, the equipment is medium- or large-scale rather than laboratory-scale and although this may hamper its use in academic research it encourages its industrial use. Experimental protocols can be designed for most cell types when only one or a few enzyme activities are of interest. Safety problems or excessive damage to the more fragile types of cells or enzyme systems are the major limiting factors. Various organic liquids can be used for permeabilization, generally those acting upon the lipid component of cell membranes. Some are already accepted for use in food processing while others do not have this clearance and so would be restricted to processes not involving edible end-products.

Permeabilization is often an almost inevitable consequence of selecting certain immobilization procedures. In particular, the use of highly reactive monomers, reactants and cross-linking reagents will lead to at least some permeabilization. The use of such immobilization methods may, in any case, be restricted to cells where loss of viability is acceptable and where the removal of unnecessary diffusion barriers is a desirable objective. Organic liquids (often simply described as 'solvents') may also bring about permeabilization during the immobilization process. They may be added specifically for this purpose or merely be the inevitable consequence of choosing a particular immobilization procedure.

A range of techniques is also available for post-immobilization permeabilization. Again, organic solvents may be used as well as the various drying procedures. Cross-linking reagents may be used even where they did not form part of the primary immobilization stage. Where the matrix is sufficiently stable it is possible to use sonication or to incubate the immobilized cell preparation under conditions causing enzymic or autolytic membrane degradation. Toxic metals, including the transition metals, can also be used to increase membrane permeability. Detergents are also a well-known, heterogeneous class of substances causing permeability increases in, or even complete disruption of, cell membranes.

It is obvious that there are many ways to achieve permeabilization and also many different degrees of permeabilization. The protocol clearly needs individual development for a specific process. There are several aspects to the benefit obtained from permeabilization. Much attention is given to the retention of cell viability, allowing the *in situ* regeneration of biocatalytic activity or increases in cell loading. This may, however, lead to cell leakage from the immobilization support and hence lead to product stream contamination. In certain applications, therefore, the absence of viable cells, but with the fullest possible retention of complex biocatalytic capability, may be desirable. In other cases, only the retention of one or two enzyme activities may be necessary and factors such as stability, cheapness and performance may be overriding. There is no doubt that the existence of an intact plasma membrane does create a significant barrier to the free diffusion of many substances. Its removal will increase mass transport and consequently allow considerably increased expressed enzyme activities. This increase may be of the order of twenty or thirty times, compared with intact cells. The benefits which can be shown for free cells as, for example, discussed by Salmon (1984), will not, of course, be as marked in immobilized preparations where the support will itself restrict mass transport.

A concomitant aspect of permeabilization is that it may reduce or eliminate undesirable biocatalytic reactions. By judicious selection of the technique, less stable systems and enzymes can be either washed out of the cells or inactivated *in situ*. This is an alternative approach to the selection or genetic modification of cells to eliminate the expression of such competing reactions, a procedure which is not always possible if the competing pathway is essential for normal cell development. Another interesting feature of the milder methods of permeabilization is the possibility of releasing certain biosynthetic pathways from the normal regulatory control operating in intact cells. Continuous outward diffusion and loss of metabolites will prevent their accumulation in cells to the concentrations which exert regulatory inhibition. Continuous 'over-production' may be achieved without the need to undertake genetical modification, as found, for example, by Murata *et al.* (1981) for glutathione production by yeast. Such methods may prove valuable in the production of currently expensive cofactors (coenzymes). In general, however, it must be stated that the loss of organic and inorganic cofactors often presents problems where complex biocatalytic conversions are being attempted. While it may be economic to add simple inorganic cofactors, as used for glucose isomerase preparations, it is very rarely economic to add expensive organic cofactors such as ATP or reduced NAD. Although cheaper substitutes have been investigated they seldom prove as efficient as natural coenzymes.

From what has been discussed already it is clear that permeabilization, known or unknown, is an aspect of almost any immobilization procedure. In the majority of published work it forms a coincidental aspect of the search for optimal biocatalyst activity and is not directly referred to in the discussion of the techniques selected or the comparative results obtained. One research area which has been particularly concerned with permeabilization is that of the production and liberation of secondary products by immobilized plant cells. Brodelius & Mosbach (1982) make specific reference to this topic, and quote results from their own laboratories (Felix, Brodelius & Mosbach, 1981). A variety of reagents can be used, ranging from organic liquids, antimicrobial agents and surface active agents to proteins. There is, of course, some doubt about the location of secondary metabolic pathways and the sites of accumulation of end products. With plant cells it may generally be assumed that the tonoplast (vacuolar membrane) is permeabilized at the same time as the plasma membrane. It is probable, however, that much of the secondary metabolism occurs in the cytoplasmic compartment, even where the products are clearly accumulating in the vacuoles of the cells. Using suspension culture cells of *Catharanthus roseus* it was shown that immobilized, permeabilized cells retain tested enzymes of the primary metabolic pathways for considerably longer than the same cells free in suspension. Initial detectable levels of the hexokinase/glucose-6-ph dehydrogense system were highest with a dimethyl sulphoxide (DMSO) treatment, closely followed by protamine sulphate, lysolecithin and poly-L-lysine. Around 25% less was detected using cytochrome C and ether and 33% less for filipin. Nystatin and acetic acid treatments gave only 50% of the activity found after DMSO treatment. It was noted that even the addition of the coenzyme NADP+ resulted in some permeabilization of untreated cells. The half-life for this two-enzyme system was about six hours for free cells and four days for immobilized cells. It is worth noting that it was not possible to demonstrate these enzyme activities in alginate-immobilized cells, due, it is suggested, to the interference of high calcium ion concentrations. Cells recovered from alginate gels were shown to possess the test activity. For the other experiments agarose was used. The *C. roseus* cells permeabilized with DMSO were shown to be competent in the conversion of exogenous cathenamine (plus reduced NADP) to ajmalicine isomers. With permeabilized cells around 90% of the alkaloids formed were found in the medium but with intact cells only 70% was excreted. The figures did show, however, that the intact cells could provide their own endogenous reduced NADP, showing nearly 90% of the activity of the DMSO-treated cells (when the latter were in a medium containing reduced NADP). It should be noted that isocitrate could substitute for added reduced NADP, showing that the DMSO-treated cells also retained the capacity to generate reduced coenzymes.

In a subsequent publication Felix & Mosbach (1982) went on to show that DMSO-treated, agarose-immobilized cells of *C. roseus* were more active than similar cells immobilized in polyurethane (hypol 3000) foam. Use of the hardening agent glutaraldehyde (GA) plus hexamethylenediamine (HMDA) almost totally inactivated the cathenamine reductase activity but had only a relatively small effect on isocitrate dehydrogenase activity except where DMSO had been used first. The use of GA + HMDA alone caused some permeabilization. Mechanical agitation of immobilized cells was shown to be deterimental to enzyme activity. Gentle handling plus a GA + HMDA treatment increased the half-life of isocitrate dehydrogenase from four days to 72 days.

Felix (1980) reviewed permeabilized microbial cells as part of his Doctoral thesis and showed (as did Gestrelius, 1983) that a wide variety of techniques have been used. It is not the purpose of the present text to review the literature exhaustively and only a few examples will be given. Kawabata & Demain (1979) studied the synthesis of pantothenic acid by *E. coli*. They studied pantothenic acid synthetase, which was produced during the exponential growth phase and showed that acetone-dried cells or a buffer extract of such cells were enzymatically active and required exogenous ATP as well as substrate. Although frozen-thawed cells were somewhat less active they were chosen for further study because of the simplicity of this procedure and the easy possibility of economically viable scale-up. They concluded that the permeabilization produced by freeze-thawing was of a subtle nature, because the target enzyme did not leak out of such cells. Interestingly, the optimum pH value for the reaction mixture was intermediate between that of the free enzyme (pH 10.0) and untreated resting cells (pH 7.0–7.5). Also, the temperature stability of the enzyme was greatest in frozen-thawed cells. The immobilization matrix used was agar and the half-life of the preparation was 100 hours at 37 °C, in the presence of the reaction mixture.

Turning to the production of reduced coenzymes, Egerer *et al.* (1982) immobilized *Alcaligenes eutrophus* in ENT-2000 and PU-6. With exogenous NAD the PU-6 immobilized cells showed a more than two-fold increase of activity over free cells and a four-fold increase over similarly immobilized, freeze-dried cells. ENT-immobilized cells were not active. Hydrogen gas was used as the reactant for the target enzyme, NAD-dependent hydrogenase. Again, a subtle permeabilization is clearly required for optimum activity. In related research Egerer & Simon (1982) immobilized *Clostridium* sp. in ENT-110 and used methylviologen as an exogenous carrier for the hydrogenation of unsaturated compounds. Before immobilization the cells were stored for up to fourteen weeks at −20 °C. The need for high glycerol concentrations in the freeze-storage medium demonstrated that some permeabilization is beneficial but total cell disruption by freezing is undesirable. The half-life of the immobilized preparation was about ten days for the hydrogenation of (E)-2-methylcinnamate to (2R)-methyl-3-phenylpropionate. In these experiments the greater stability of methylviologen compared to NAD is used to benefit. The system was stated generally to have at least an order of magnitude better activity than most other comparable preparations and to exhibit activity against a wide range of substrates.

In their work with the thermoacidophilic bacterium *Calderiella acidophila*, De Rosa *et al.* (1980) found polyacrylamide entrapment gave an almost nine-fold increase in detectable β-galactosidase compared to intact cells. This effect was attributed to permeabilization and appeared to be 75% of that obtained using acetone-treated free cells. Acetone treatment of the immobilized cells did not produce any greater activity, so it may be assumed that the permeabilization produced by the immobilization is complete and that diffusional limitations or enzyme inactivation are responsible for the observed decrease in activity. From a practical viewpoint, however, the more important finding was that the activity of the immobilized cells was not reduced by acetone washing. This opens the way for using organic solvents to remove accumulated fatty materials from columns used to treat mixed component streams. In the present example the target process was the treatment of milk products to hydrolyse lactose. The use of a thermophilic bacterium, of course, confers the advantage of genetically-determined heat stability upon the target enzyme. Oshima (1978) discussed some comparative aspects of heat-stable enzymes from thermophilic micro-organisms.

An interesting method for the release of metabolites normally retained in cells was described by Reed *et al.* (1986). The organisms involved were the cyanobacteria *Synechocystis* and *Synechococcus* both free and immobilized in alginate. Reduction of the salt content of the medium promoted release of free amino acids and carbohydrates and the treatment could be repeated for several cycles. This method might well prove applicable to other types of cell.

Protoplasts

The act of removing the cell wall from a plant or microbial cell causes the protoplast to become as sensitive to osmotic swelling as an animal cell. In solutions of low osmolarity a protoplast will swell up and burst. One immediate benefit of immobilization in alginate gels, which was noted by Scheurich *et al.* (1980), was that such protoplasts did not burst because the matrix acted as a pseudo cell wall. They could not detect any morphological or permeability changes in protoplasts of *Vicia faba*, even after 14 days in the immobilized state. In a suitable growth medium new cell walls are synthesized, just as they are by free protoplasts.

Several papers delivered in 1983 at the 6th International Protoplast Symposium dealt with the immobilization of protoplasts. Dix *et al.* (1983) drew attention to the benefits of increased secondary product release by immobilized plant cells and the restriction of growth which channels metabolism towards secondary products rather than to cell division processes. Their protoplasts from *Nicotiana plumbaginifolia*, under suitable conditions, regenerated cell walls and retained high viability. When growth was restricted in an auxin-free medium the cells remained viable for many weeks but did not rapidly disrupt the beads by growth, as did cells in an auxin-containing medium. Linse & Brodelius (1983) immobilized protoplasts of *Daucus carota*, with a method involving the formation of an emulsion of the aqueous phase in vegetable oil, using carrageenan or agarose as the matrix. The immobilized protoplasts showed considerably better retention of activity than free protoplasts during batch reuse. The target hydroxylation reaction was the conversion of digitoxigenin to periplogenin. The better performance of the immobilized protoplasts was attributed to improved resistance to mechanical damage.

Larkin (1981) reviewed the agglutination and immobilization of plant protoplasts, with particular attention to the cell surface recognition sites and their interactions with binding molecules. In comparison with animal cell studies this is an under-researched subject, despite the fact that the lectins used in so many animal cell studies are derived from plant material. Most plant research has centred around phytopathological studies of host–pathogen interaction. There is, however, a modest number of references to the agglutination of plant protoplasts by lectins, although as Larkin pointed out this may be totally unrelated to any *in vivo* action of the lectins in the plant species which produce them. Describing his own experiments, Larkin reported tenacious binding of plant protoplasts to lectins covalently linked to either glutaraldehyde cross-linked serum albumin sponges, or collagen membranes. The review is, however, not in any way concerned with the biotechnological uses of immobilized cells.

Microcarrier beads, such as the beaded, cross-linked dextrans, are routinely used for the culture of animal cells. Bornman, Olesen & Zachrisson (1983) coated these beads with wheat germ agglutinin and showed a three-fold increase in the binding of *Beta vulgaris* protoplasts. Microscope studies revealed up to forty protoplasts attached to a single carrier bead. Many biotechnologically interesting uses of microcarrier bound protoplasts are possible but it is likely that they would be too expensive for the industrial production of secondary products.

Among the scientific benefits of immobilizing protoplasts in gels are the greater ease of culture of delicate protoplasts, due to protoplast stabilization and also the improved success rates for protoplast fusion. As examples of this, Shillito, Paszkowski & Potrykus (1983) used agarose for the immobilization of plant protoplasts and Vidoli *et al.* (1982) used calcium alginate for *S. cerevisiae* protoplasts.

Organelles

The immobilization of organelles can be clearly separated from that of protoplasts because the former never have the potential to regenerate into intact cells. Organelles can be considered to be multifunctional biocatalysts, intermediate in complexity between viable whole cells and immobilized enzymes. Almost all are membrane-bounded or membrane-containing preparations and will contain enzymes and complex enzyme systems already self-immobilized and orientated in these membranes. Tanaka & Fukui (1983) reviewed the state of research on immobilized organelles.

Many experiments have been carried out on chloroplasts. These are easily obtainable by relatively simple extraction and purification procedures, starting with cheap, readily available fresh plant material. Unless special steps are taken during the cell disruption and chloroplast recovery stage, the chloroplasts are almost certain to have lost the outer of the two concentric membranes which surround them. At the same time, in aqueous media, a large proportion of the soluble enzymes involved in the fixation of carbon dioxide will also generally have been lost. Immobilized chloroplasts are not, therefore, seen as an alternative to whole cell or whole organism photosynthesis to create organic compounds *de novo*. Rather, they are used as photoconverters of light energy into chemical potential energy. This may take the form of active forms of coenzymes such as ATP or $NADPH_2$ which can be continuously regenerated by light in complex immobilized biocatalytic systems, thus replacing the need for expensive, exogenous coenzymes. Another alternative which has been extensively investigated, but which so far is essentially of academic rather than industrial significance, is the generation of hydrogen by the photo-chemical conversion of water, in the presence of a suitable hydrogenase.

Whatever system is being investigated there is no doubt that immobilized chloroplasts are more stable than free chloroplasts. A wide range of immobilization techniques has been used. Because the electron transport system resides in the internal thylakoid membrane system it is possible to use relatively harsh methods, such as glutaraldehyde cross-linking (fixation), acting on the remaining outer chloroplast membrane. The activity of photosystems I and II has been demonstrated in such preparations, as well as oxygen evolution. Although spinach is the classical material for such studies, other species, such as lettuce and tobacco, have also been used. Many studies have also been carried out with entrapment in a wide variety of gels. Agar, calcium alginate, polyacrylamide, polyvinyl alcohol, collagen, glutaraldehyde cross-linked albumin, pre-polymer resin and urethane systems and a few others have all been tried. Papageorgiou (1983) immobilized spinach chloroplasts in bovine serum albumin using glutaraldehyde and used freezing to permeabilize the resulting preparation. In some systems added natural or synthetic redox carriers have been used as well as various microbial hydrogenases. The ability of chloroplasts to generate oxygen poses some problems when using oxygen-sensitive hydrogenase systems. As early as 1977 Kierstan & Bucke demonstrated the ability of calcium alginate-immobilized spinach chloroplasts to reduce DICPIP (dichlorophenol-indophenol) when illuminated. This simple modification of a classical experiment is an interesting teaching system on immobilized organelles.

As might be expected, the physical characteristics of the immobilization system influence the activity of the final preparation. Gisby & Hall (1980), for example, found that calcium alginate films held on supporting grids gave higher functional activity than beads. Their work is of particular interest because the synthetic co-catalysts methylviologen and platinum-polyvinyl alcohol replaced natural (and more unstable and expensive) ferredoxin and hydrogenase enzyme. The direct generation of an electrical current by immobilized chloroplasts (using a polyvinyl alcohol film on a tin oxide coated electrode, all covered with a selectively permeable membrane) has been shown by Ochiai *et al.* (1979). However, it is clear that the current generated was considerably less than that required for the 60 000 lux of light used to produce the electrode current. The same research group (Ochiai *et al.*, 1980) has also immobilized living cells of the thermophilic blue-green bacterium *Mastigocladus laminosus* on a similar electrode to produce a photocurrent. The reader is reminded that chloroplasts are generally considered to be a highly advanced symbiotic form of blue-green bacteria.

When non-oxygen evolving photosynthetic bacteria are disrupted a preparation of their light-transducing membranes can be made. These are comparable in functional activity to the thylakoid membranes within chloroplasts and are known as chromatophores. Paul & Vignais (1980), for example, immobilized chromatophores from *Rhodopseudomonas capsulata* in calcium, barium and strontium alginate gels. The barium-type beads had the best physico-chemical properties and had a high photophosphorylation activity. As expected, the apparent Michaelis constants were increased by entrapment. In the present example it should be remembered that calcium alginate gels are rather unstable in the concentrations of inorganic phosphate needed to ensure adequate synthesis of ATP. Although most work has probably been done with *R.*

capsulata, Tanaka & Fukui (1983) briefly review results from several other species of photosynthetic bacteria. Interest has centred upon ATP generation or electrode potential generation systems and entrapment has been tried in various other gels, such as glutaraldehyde cross-linked proteins, polyacrylamide and synthetic pre-polymer systems. In general, chromatophore preparations are more stable than chloroplasts and their half-life is considerably improved by immobilization, while still retaining a considerable capacity to generate ATP.

Although their preparation is somewhat more complicated, there have been a number of studies on immobilized mitochondria. Arkles & Brinigar (1975) found rat liver mitochondria to adhere readily (at 27 °C) to glass beads coated with various alkylsilanes. This self-adsorption system was reversed at low temperatures (0 °C) allowing defunct mitochondria to be readily replaced by more active ones. Most experiments with mitochondria have been concerned with the study of the properties of the mitochondria themselves but they offer a potential system for the generation of ATP or certain metabolites from their internal metabolic pathways. Godbole *et al.* (1983), for example, studied the production of L-malic acid by rat liver mitochondria immobilized in polyacrylamide gels which had been polymerized using gamma irradiation. Maximum fumarase activity was found with 16% acrylamide and 0.8% *N,N'*-methylene-bis-acrylamide entrapping 6% mitochondria preparations (all w/v figures) in the presence of 0.1 M phosphate buffer of pH 7.5. As expected, the immobilized mitochondria showed enhanced temperature stability and could be repeatedly used for batch conversion of fumarate to L-malic acid. Various methods were tried to suppress succinic acid formation, of which malonate plus sodium dodecyl sulphate was the most successful. Yields of malic acid were only slightly above 1% and there clearly needs to be some further improvement in the system if it is ever to reach a commercially viable state. Subfractions of disrupted mitochondria have also been immobilized and their biocatalytic activity investigated.

Various other organelle fractions, such as peroxisomes and microsomes have also been immobilized. Many of these studies have been largely directed towards an elucidation of their *in vivo* activities but some possible commercial applications could arise. Ahern, Kator & Sada (1983) for example, investigated prostaglandin synthesis by ram seminal microsomes.

Co-immobilization

The term co-immobilization can be widely defined to include all permutations of the three components, enzymes, organelles and cells. The interactions between living micro-organisms are well documented

Table 6.2. *Some examples of defined co-immobilization systems*

Cells with Cells
  Saccharomyces cerevisiae + *Escherichia coli*
  Saccharomyces cerevisiae + *Bacillus ammoniagenes*
  Chlorella vulgaris + *Providencia* sp.
  Klebsiella pneumoniae + Rhodospirillum rubrum

Cells with Enzymes
  Aspergillus niger + catalase
  Saccharomyces cerevisiae + protease
  Saccharomyces cerevisiae + β-galactosidase
  Escherichia coli + alcohol dehydrogenase

Organelles with Enzymes
  Chloroplasts + hydrogenase
  Chromatophores + hexokinase

and it is as simple to immobilize a mixed culture as it is a pure culture. Wastewater treatment plants are invariably mixed-culture systems involving natural adsorption to surfaces or flocculations and specifically promoted encouragement of these processes. Some interesting results have also been obtained with the more refined methods of immobilization, such as gel entrapment. Although there is the inevitable loss of activity during the immobilization process there may be useful gains. Methanogenic bacteria, for example, may be protected from oxygen by the diffusion barrier created by gel immobilization in agar, together with an increase in methane formation per cell as shown by Karube, Kuriyama, Matsunaga & Suzuki (1980). The characterization of such undefined mixed cultures poses considerable problems to the researcher and reports are largely confined to the very extensive literature arising from the waste treatment and fermented food industries. In both of these the economic climate is generally such as to discourage most innovative procedures which involve additional and unproven stages.

The situation with defined co-immobilization (Table 6.2) is rather different because here the organisms can be selected to show complementary biocatalytic activities. Chibata & Tosa (1981), for example, draw attention to the benefits of being able to balance the respective enzyme activities of two different species to obtain improved yields, quoting from their own published work on glutathione production by *Saccharomyces cerevisiae* with *Escherichia coli* and NADP production by *S. cerevisiae* and *Brevibacterium ammoniagenes*. In these examples a yeast and a bacterium are involved. Wilkstrom *et al*. (1982) co-immobilized the alga *Chlorella vulgaris* with a bacterial *Providencia* sp. An interesting, but at this stage small-scale, illuminated columnar reactor was used, packed with agarose-entrapped cells. The algae acted mainly as an internal generator of oxygen to the bacteria which carried out the target production of keto acids using their amino acid oxidase system. A 10-fold increase in conversion was obtained using leucine as the substrate.

Certain inevitable problems do, however, arise when using defined co-cultures. While natural mixed culture situations are inherently stable and self-balancing (allowing for cyclic variations in the actual numbers of the individuals of each species present) it is necessary to set up a defined co-culture with a particular ratio of numbers between the different species present. Normally this is done by an empirical experiment of mixing various proportions and seeing which performs best. The snag to this, however, is that the stability of the discrete target activities in the two (or more) species may not be the same. Whatever operating conditions are chosen they are unlikely to be optimal for all species present. Furthermore, if viable cells are present the growth rates, under the chosen conditions, are also likely to favour one species. The result of both of these is a change away from the selected optimum ratio of activities. This may be empirically overcome by having an excess of the more sensitive biocatalyst present. This may, indeed, be the situation chosen empirically from the outset of the study, if the criterion used was the long-term stability of the system rather than the initial rate of bioconversion. One possible way round this is to immobilize the component species in separate gels and, effectively, to use a two-stage reactor, the individual stages being separately optimized. This solution, however, may not provide a sufficiently close proximity of the two species to overcome problems of product inhibition in the first stage. Another approach might be to use two different immobilization methods with the more labile biocatalyst in, say, calcium alginate, which could be redissolved when the activity became too low and then replaced by fresh biocatalyst. Some workers put their faith in the use of inherently stable biocatalysts, such as those present in thermophilic species, where comparable stabilities might be expected among different species.

Because gel immobilization inevitably restricts oxygen diffusion it may prove of great benefit in protecting oxygen-sensitive biocatalysts from exogenous oxygen. Weetall, Sharma & Detar (1981), for example, used agar to co-immobilize *Klebsiella pneumoniae* and the photosynthetic bacterium *Rhodospirillum rubrum*. This system was claimed to give a very high yield of hydrogen when fed with glucose.

Instead of co-immobilizing two species together it is possible to co-immobilize an enzyme and a micro-organism. Hartmeier & Doppner (1983) permeabilized mycelia of *Aspergillus niger*, containing glucose oxidase and catalase, with isopropanol. Glutaraldehyde was then used to co-immobilize concentrated catalase and egg albumen to the mycelial surface. Unlike most immobilized preparations the apparent Michaelis constant of the glucose oxidase was actually decreased. The target reaction for their system was conversion of glucose to gluconic acid. The co-immobilized preparation could use exogenous hydrogen peroxide as the source of oxygen, without the usual problem of rapid enzyme inactivation. This benefit was attributed to the excess catalase bound to the mycelia. A rather similar approach was taken by Hartmeier (1981) who adsorbed a proteolytic enzyme onto the surface of *Saccharomyces cerevisiae* and then immobilized it permanently by cross-linking with glutardialdehyde in a tannin solution. This co-immobilized preparation was used to produce wine must with lowered foaming and slightly increased fermentation rates, the former indicative of protein degradation and the latter possibly due to a consequential increase in available nitrogen.

An interesting form of co-immobilization was used by Hahn-Hagerdal (1981). β-galactosidase was first covalently coupled to alginate, using carbodiimide. Conventional calcium alginate immobilization of this with *S. cerevisiae* was then used to obtain a system capable of fermenting the lactose from milk whey to ethanol. This, once again, clearly demonstrates the way in which immobilization can be an alternative to a recombinant DNA approach to genetic insufficiency. In an interesting review of the whole subject of co-immobilization Hahn-Hagerdal (1983) also quotes both published and previously unpublished results of β-glucosidase immobilized with *S. cerevisiae* to bioconvert cellobiose to ethanol.

Another approach in enzyme/cell co-immobilization is to use the cell as a regeneration system for active forms of coenzymes required for the catalytic activity of the enzyme. Burstein *et al.* (1981), for example, used the respiratory electron transport chain of *E. coli* to regenerate the oxidized form of NAD, allowing the continued conversion of ethanol to acetaldehyde by alcohol dehydrogenase co-immobilized with the bacteria and serum albumin using glutaraldehyde. Clearly many similar systems can be devised. The use of prokaryotes as the coenzyme regeneration part of the system is much easier than the comparable use of organelles such as chloroplasts and mitochondria or isolated membrane fragments. A few researchers have attempted the co-immobilization of organelles and micro-organisms but this does not appear to be a very promising area because of the inherent instability of the organelles, which will always prove to be the weak link in the system. With all co-immobilizations, it is the least stable component which determines overall stability. Cocquempot *et al.* (1982) co-immobilized spinach chloroplast thylakoid membranes with hydrogenase from *Desulfovibrio gigas*. Calcium alginate or glutaraldehyde plus inert protein methods were used for the membranes and covalent linkage of the hydrogenase to porous silica or dextran beads, which were then incorporated with the cells during their immobilization. Close proximity of membranes and hydrogenase was necessary for optimal hydrogen production. The major interest of this method resides in the dual immobilization procedures used, allowing the best method to be used for both the enzyme and the more complex membrane systems. Similar methods have been used to co-immobilize hexokinase and an ATP-regeneration system consisting of bacterial chromatophores.

Animal cells

The study of the self-immobilization of animal cells in culture is sufficiently extensive to justify a book in its own right. It follows, therefore, that this very small section in the present book can only briefly describe the salient features, from a comparative viewpoint. As discussed in Chapter 1, animal cells in culture can be divided into two types, firstly, those that proliferate freely, even in suspension culture and, secondly, those described as anchorage-dependent. The latter will only show cell division after they have adhered to a solid surface. It is both expensive and technically difficult to culture animal cells and they are easily damaged, because they lack a protective cell wall around the plasma membrane. It is reasonable that only the most mild immobilization methods have been studied and that complex biosynthetic or biotransformation reactions are of interest rather than the single or few-step biocatalytic activities which form such a major concern of those immobilizing microbial cells. The adhesion of animal cells to the surface of glass or plastic culture vessels need not concern us here. The small quantities of cells which can be produced by such techniques are more of academic research interest than of potential commercial application. Attention focuses, therefore, on the various microcarrier systems which have been developed, consisting of small, spherical supports to allow a pseudo-suspension culture of anchorage-dependent cells.

In the initial experiments of Van Wezel (1967) DEAE-Sephadex A-50 (a registered trademark of Pharmacia Fine Chemicals), sold as an ion-exchange gel, in bead form, was used. These beads possess a high surface to volume ratio, as well as having a specifically charged surface and a suitable density to allow easy suspension in aqueous liquids without the need for too vigorous (and hence damaging) stirring. Another feature which they possess is their transparency to visible light, allowing the direct microscopic examination of the cells. As discussed by Hirtenstein & Clark (1983), the degree of substitution of the support is important in obtaining optimal culture conditions and this led to the development of improved carriers. Hirtenstein, Clark & Lindgren (1980) reviewed the various carrier matrices available at that time and concluded that a dextran-based support was best. By restricting the charge groups to the outer surface of the beads, using appropriate manufacturing techniques, it is possible to avoid the detrimental effects of unwanted ion-exchange binding of media components and/or cell products. The trade literature produced by Pharmacia Fine Chemicals (1981) provides an excellent account of the principles and methods of microcarrier animal cell culture. The Cytodex (registered trade mark of Pharmacia Fine Chemicals) range of carriers includes those with homogeneous and surface-only N,N'-diethylamino ethyl groups as well as a type coated with covalently bound denatured collagen. Each type has its specific benefits and uses. Hirtenstein & Clark (1983) list over a hundred cell types which have been cultured on Cytodex carriers. The major problem with these carriers is that they are capable of only very limited (if at all) reuse.

Schulz, Krafft & Lehmann (1986) described the use of dextran microcarriers substituted with DEAE residues, available with various nitrogen contents. A recombinant mouse-L tk$^-$ cell line, carrying human B-interferon gene was used as the test system, grown in a bubble-free glass fermenter. Good cell attachment was obtained.

Varani *et al.* (1983) described the properties of a new glass bead microcarrier and compared it with available plastic and DEAE-dextran microcarriers. Previously, glass has not found favour as a carrier because of its high density. The new beads have a density of only 1.04 g/l and a diameter of 100–150 μm. Three human cell lines were used for the study. For these the growth on the glass beads was comparable with that on the already commercially available carriers and the cells could be more easily recovered by trypsin treatment than they could from the dextran-based microcarriers. It was shown that the glass beads could be reused, giving comparable yields to the original first-use beads. Examination of the attached cells under the microscope clearly showed a difference in morphology. With the glass and plastic Biosilon (trademark of A/S Nunc) supports, cells were attached to the surface by long filopodia. On Cytodex 1 and Superbeads (trademark of Flow Laboratories) the cells were more closely attached, apparently by the whole edge of the cell and not by filopodia. The extent of bridging between beads and multiple layer formation on the bead surfaces was also greater on the glass beads. These differences were related to ease of cell removal by trypsin and it was also suggested that metabolic activity might be influenced by the type of microcarrier used.

Although extensive information is available for DEAE dextran-based microcarriers there is still a great need for further information on the factors which affect cell attachment, growth and metabolism on microcarriers. Reuveny *et al.* (1983) set out to develop a system which allowed flexibility in the properties of the microcarrier, basing their studies on polyacrylamide beans which they prepared and derivatized themselves. Primary or tertiary amino groups were introduced throughout the gel, in various quantities, to give a wide range of exchange capacities. Four animal cell lines, with distinct cellular morphologies, were studied. Derivatization with primary amines gave faster cell attachment and at lower exchange capacities than when tertiary amines were used. Epithelial-type cells grew better on primary amino-derivatized beads. Cells with fibroblast morphology grew better on beads derivatized with tertiary amino groups. The authors see their studies as possibly leading to microcarriers with specific properties suitable for particular cell lines, because of the ease with which the properties of their polyacrylamide beads can be modified.

There is no doubt that much more, experimentally demanding, research needs to be done on carrier systems for animal cells. The increasing use of cultured animal cells (as discussed, for example, by Spier, 1980), is likely to lead to the development of new and improved techniques of animal cell immobilization. This will be yet another area of biotechnology where immobilized cell technology offers an alternative to recombinant DNA technology.

7   Reactor design and operation

The choice of an appropriate design of immobilized cell reactor is fundamental to the success of a process. In general, the type of reactor chosen determines the type of immobilized cell preparation which can be used in it and vice versa. Some immobilized cell preparations may only be usable with a single type of reactor while others, such as inorganic, porous, spherical beads, may be satisfactory in diverse types of reactor. The matching of the two components depends on a proper understanding of the conditions which exist in the various types of reactor and the physical and biological properties of the immobilized cell preparation. The latter has been discussed in earlier chapters.

There are two major ways of expressing the productivity of an immobilized cell reactor. The volumetric (or reactor) productivity expresses the amount of product formed in terms of unit volume and unit time. The specific productivity is expressed per unit mass of immobilized cell preparation (or of actual biomass) and unit time. For reasons explained in Chapter 2 it is not easy to determine the amount of biocatalytically active biomass which is present in an immobilized cell reactor. It is also found that the amount present may vary, due to loss of biocatalytic activity or cell growth increasing the amount of biocatalyst present. For these reasons many biochemical engineers consider that only the volumetric productivity is of any real significance. Even where the reactor volume and substrate flow rate are kept constant, the changes in the activity of the immobilized cells will result in time-related changes in productivity for any continuous-flow system. Operational strategies need to take account of such inevitable changes, bearing in mind the nature of the immobilized cell preparation.

By its very nature a batch process gives rise to a changing environment within an immobilized cell reactor. At the start the substrate concentration will naturally be high and that of the product low and vice versa at the end. Even in a completely mixed reactor this will result in a progressive fall in the rate of formation of the target product. At any one instant of time the product concentration would be expected to be the same in all parts of the reactor. The profile of the reduction in the rate of product formation will obviously depend on the nature and activity of the immobilized cells and the substrate/product concentrations. It is important here to distinguish immobilized cell reactors, operating under

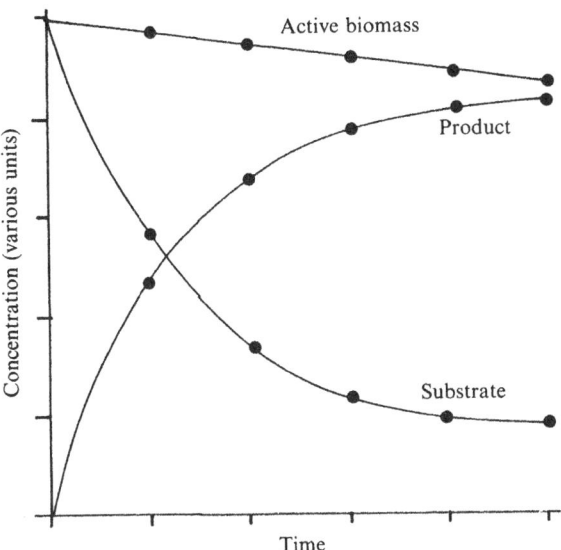

Figure 7.1. Graphical representation of the changes in the concentrations of substrate, product and active biomass with time in a batch reactor containing a non-viable immobilized cell preparation.

conditions which allow viable immobilized cells to grow, from those operating under non-growth conditions or with non-viable cell preparations (Figure 7.1). There is little difference between the former reactor and an ordinary fermenter. For immobilized cell reactors allowing natural cell immobilization onto surface supports, just as in a batch fermenter, there will be an initial period of growth and establishment. The major difference would be expected to be the greater life-span of the immobilized cells compared with free cells, which generally show a decline in activity relatively soon after reaching the stationary phase of growth. This, of course, is the period when valuable non-growth-related products may be formed. Cells are frequently immobilized to extend this production phase.

Although many industrial fermentations are run in the batch mode there are many advantages in operating a continuous reactor. By its very nature, a batch process has extensive non-productive phases between batches, when the reactor is being emptied, cleaned, refilled and sterilized. During continuous operation the immobilized cells can be kept in a relatively constant environment, compared to a batch process where, at the very least, substrate and product concentrations will be changing. This environmental constancy also makes it easier to control the reactor within desirable operational limits to give a constant productivity. Provided the cells can be kept in the productive phase, there will be a constant outflow of product which can then be recovered by continuous downstream processing. The equipment used can be smaller for any given production target because, for example, a whole week's product does not have to be processed in the short time period which is often dictated by unstable bioproducts. Greater efficiency in the use of all back-up services is also obtained. Although the economics of production will naturally vary from place to place and process to process, it is generally true that a continuous production system has economic advantages.

Both batch and continuous reactors can be operated as completely mixed vessels. This implies that, within practical limits, all parts of the fluid within the reactor have the same composition. In a batch reactor all the contents have the same fixed residence time within the reactor. In a continuous reactor, with complete mixing, it follows that substrate molecules entering the reactor may appear almost immediately at the outflow. Other molecules will remain in the reactor for longer periods and there will clearly be an average residence time and a particular residence time distribution which will depend on the characteristics of the reactor vessel. Vessels may need to be larger than might be expected on the basis of calculated product output. It may be necessary to distinguish between the concentration distribution, which depends partly on the biocatalytic activity in the reactor as well as on the mixing, diffusional forces and fluid mechanics of the system, and the residence time distribution. The residence time distribution can be easily determined experimentally by the addition of an inert tracer substance. Because reactor systems have the same residence time distributions it is not necessarily true that they have the same performance, particularly where comparison is made between single and multiple vessel systems.

At the opposite end of the mixing spectrum from the completely mixed continuous reactor is the theoretical ideal plug-flow system in which the deliberate creation of a gradient from inlet to outlet is attempted. This is most easily visualized as a tubular reactor in which diameter is small in comparison to length. If no mixing took place every molecule entering would have a fixed residence time before it emerged at the outflow. During this time biocatalytic activity would be taking place and a concentration gradient (Figure 7.2) would be established between inlet and outlet, with the maximum concentration of substrate at the inlet and of product at the outlet. This is the type of flow sought after by biochemists for certain types of analytical system and can be reasonably achieved in tubes of relatively narrow bore, with high flow rates and minimal flow obstructions. In immobilized cell reactors all manner of unwanted mixing, diffusion and fluid mechanics effects prevent the creation of a mathematically ideal plug-flow system, but the obvious benefits for certain processes mean that plug-flow reactors are important, even when their performance is not absolutely in accord with ideal

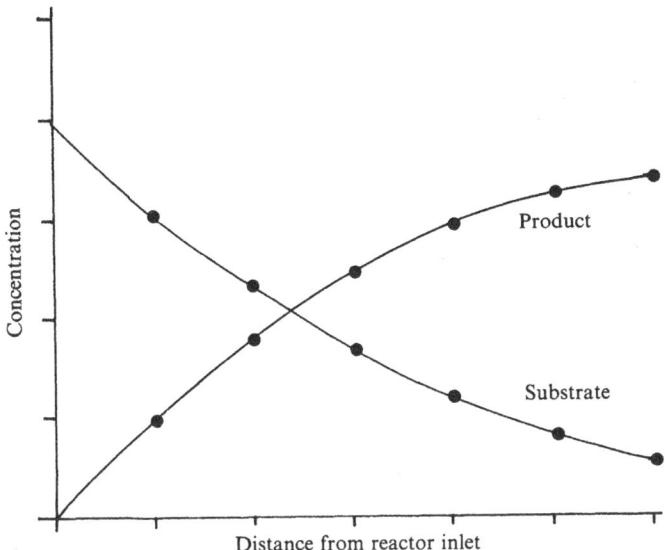

Figure 7.2. Graphical representation of the concentration gradients of substrate and product in a plug-flow reactor.

plug-flow kinetics. Forster *et al.* (1982) described the use of a lithium tracer to investigate mixing characteristics in a fixed film reactor. The use of a non-metabolized tracer is probably preferable to the usual method of tracing the path and dispersion of a potential nutrient. It is worth mentioning here that, although the terms homogeneous and heterogeneous could be applied, respectively, to completely mixed and plug-flow reactors, it is preferable to reserve their use to that of the classical chemical terminology. In this, homogeneous implies that only one phase (for example liquid) is present, while heterogeneous indicates that two or more (such as solid, liquid and gas) are present. Other modes of operation of reactors are possible, for example, fed-batch and the transient regime of Pickett, Topiwala & Bazin (1979). These do not appear to have been widely used for immobilized cell reactors.

Before proceeding to a consideration of actual reactors used for immobilized cells, it is helpful to summarize the main characteristics of each type. If it is required that the immobilized cell preparation itself is capable of movement within the reactor, three major possibilities are available. In the stirred tank reactor an impeller is used to provide the motive force mechanically. In the air- (or gas-) lift reactor the reduction in bulk density of the liquid, when gas bubbles are entrained in the liquid, is used to cause the mass flow of the liquid upwards in one part of the vessel. At the top the gas bubbles escape and denser liquid returns in a

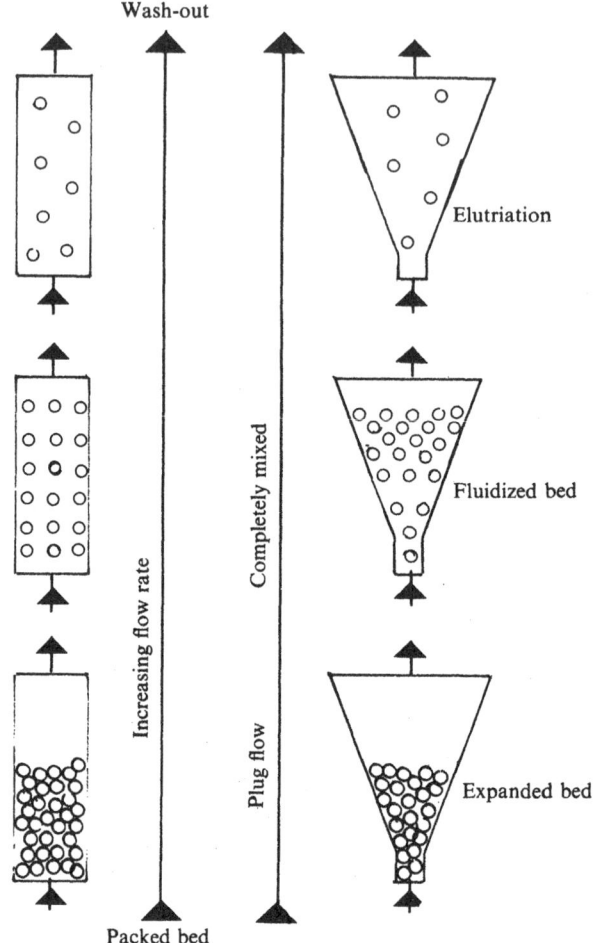

Figure 7.3. Diagram illustrating the relationship between flow rate and reactor characteristics for columnar and inverted cone vessels. The latter will show less elutriation due to reduced flow rates in the expanded upper portion.

separate part of the reactor. Generally the system used consists of concentric tubes, with the inner shorter one providing the upflow by appropriate positioning of the gas sparger. It is possible to operate with an external loop upflow. The external loop system generally results in greater shear damage as the immobilized cell preparation enters and leaves the external loop. The major benefit of the gas lift comes from its lower shear stress and good aeration properties, particularly when tall vessels are used. The rate of mixing can be separately controlled from the bulk rate of liquid flow in and out of the reactor, when operating in a continuous mode.

An alternative approach is to use the mass flow of liquid through the reactor to suspend the immobilized cell preparation, without aeration, in a single compartment vessel which might be columnar or inverted cone-shaped. This, depending on the extent to which the individual immobilized cell preparation support particles move, may be described as an expanded-bed or a fluidized-bed reactor (Figure 7.3). The density of the support particles must clearly be matched to the mass flow rate of the liquid. At the top of the vessel a settling zone may be required, as with gas-lift reactors, to prevent loss of the immobilized cells through the outlet. The term tower reactor is often used to describe a particular shape of vessel of, say, at least 10 times greater height than breadth. Tower reactors may be essentially fluidized beds or may show some additional upflow due to gas input from a sparger at the base, when they are described as bubble columns. Immobilized cell preparations may also generate gas bubbles and these may prove helpful or detrimental, depending on the circumstances. In theory, the fluidized bed could be maintained by an external pump recirculation which is separate from the overall flows of liquid through the reactor. It will be obvious that mixing characteristics of fluidized-bed reactors can be complex, with elements of both plug-flow and completely mixed characteristics.

With organisms that are filamentous or naturally form films on surfaces, it is possible to use a different strategy for the reactor design, particularly where it is necessary to ensure that the cells in the film are adequately aerated. This has led to the development of the rotating disc and rotating cylinder reactors (or fermenters if the initial growth phase of the film is considered). The support discs or cylinder (Figure 7.4), are rotated in a trough partly filled with the growth medium, so that the cells are alternately exposed and submerged. Speed of rotation is usually relatively slow. When adequate biomass has developed the immobilized cells can be used to carry out the desired biocatalytic reaction, either in batch or continuous mode. In theory the discs or cylinders could be prepared with an artificially immobilized cell film on them. In practice, this does not seem to be a reactor type which has been considered by many researchers to have sufficient benefits to justify its construction.

Selectively permeable membranes have been used to divide reactors, either internally in one vessel or externally as part of a recycle loop, into separate compartments. Similar systems have been explored with fermenters, where the major problem is direct microbial growth on the membrane surface or blockage of the pores in the membrane by microbial metabolites. The major benefit of using separately immobilized cells with a membrane reactor is clearly that it prevents microbial growth on the membrane surface. The major drawback is the additional diffusion limitation factor arising from the membrane. Membranes can be made of a wide range of materials, some of which can be steam sterilized, and with

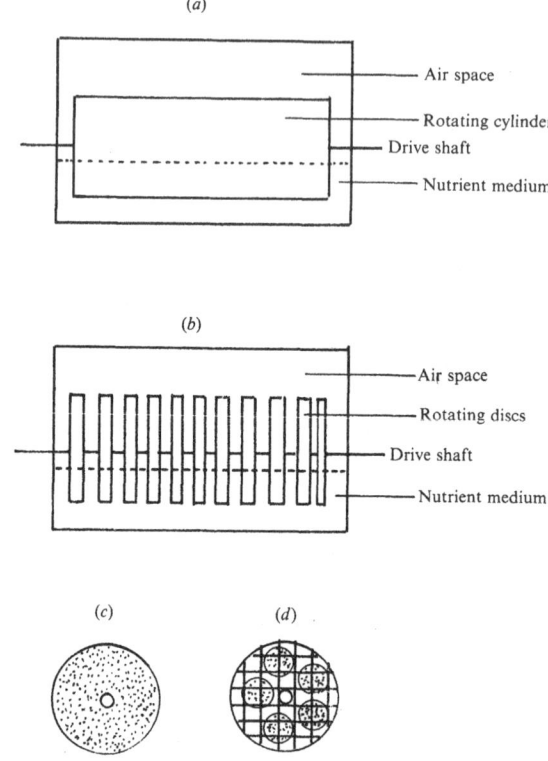

Figure 7.4. Diagrammatic representation of some horizontal immobilized cell reactors. (*a*) Rotating cylinder reactor; (*b*) rotating disc reactor; (*c*) surface growth on a solid disc; (*d*) gel-entrapped cells held in mesh discs.

a variety of different average pore sizes. Dialysis membranes have very small pores and retain all but the smallest organic molecules. Ultrafiltration membranes can be obtained with fairly precise molecular mass cut-off points, in the macromolecular size range of proteins. They have been extensively used in enzyme reactors as a simple means of immobilizing free enzyme into one compartment of the reactor. Obviously the relative size and configuration of the two compartments is crucial to the success of such systems and a particular system, with high membrane surface to reactor volume ratio, called a hollow-fibre reactor, has been developed. Although, in theory, either compartment could be used to immobilize the cells, this is usually the outer compartment of a hollow fibre device. The membranes may be isotropic (symmetric) with no difference between the two sides, or anisotropic (asymmetric) with the effective pore size on one side and the other side having larger feeder pores. These help to prevent blockage of the pores during use, provided the flow is in the correct direction. Membranes of the type used for sterile filtration, with much larger pores of the order of size of bacterial cells, could be used in membrane reactors for immobilized cells. These do not appear to have been extensively used, probably because the membranes are not readily available in convenient forms. Using membranes to retain immobilized cell preparations (such as gel-entrapped cells) may have advantages for some high value products, but simpler types of reactor are generally more cost-effective.

In many cases it has proved satisfactory to have the immobilized cell reactor packed in such a way that only the fluid phase moves. This may take the form of a simple packed-bed reactor, with relatively isodiametric (often spherical) immobilized cell preparations. In these the flow may be either upwards or downwards, the direction chosen being related mostly to the mechanical properties of the cell support material. The length to diameter ratio can vary within quite wide limits, depending partly on the extent to which plug-flow conditions are required. Instead of packing with discrete support particles it is possible to use sheet or honeycomb forms of cell support. Cells may either be self-immobilized during reactor start-up or pre-immobilized in or on the support material. Such reactors may show good hydrodynamic characteristics and be little affected by particulate matter suspended in the fluid or cell overgrowth during operation. A simple tubular reactor will have no cell support material other than the walls of the tube itself. For some specialized uses this may be a convenient form of reactor, but tube diameter generally has to be kept small to ensure adequate contact between substrate and immobilized cells.

For columns with non-deformable support materials and non-particulate input streams, down-flow operation may be entirely satisfactory, but upflow is frequently used to prevent bed compaction and concentration of particular matter at the outflow end of a column. Where there is heavy contamination of input streams by fine particulate organic matter, a down-flow (trickle filter) system may be satisfactory, with a deliberate mechanism of partial retention of organic particulates to allow their fuller degradation. Gravity-assisted flow is obviously cheaper to operate and economic factors may be of significance in using such systems to treat high-volume, low concentration, low-value process streams, such as domestic or industrial waste waters.

Types of reactor

Mobile support systems

Stirred-tank reactors. The most commonly available reactor type is the stirred-tank reactor (STR). This attempts to achieve completely mixed conditions within the reactor by the use of an impeller system, generally mounted on a rotating shaft, which may enter from either the top or the bottom of the vessel. With small vessels the impeller system may be driven by a magnetic coupling rather than a direct drive shaft. STRs mixed by simple stirrer bars on the base of the vessel are generally unsuitable for use as immobilized cell reactors, due to the milling effect that the rotating bar has upon particles which are drawn under it. This could be overcome by appropriate design modifications but the effort involved is not generally considered worthwhile. Depending on the dimensions of the vessel and the amount of mixing required, the impeller system can be varied within wide limits. The number, shape and position of the blades can be modified and these have been the subjects of extensive study by biochemical engineers. Numerous books have been published which deal with the biochemical engineering aspects of STRs and, indeed, the many other types of reactor which are available. For expansion on the brief account which is provided here the reader is referred to such books as Aiba, Humphrey & Millis (1973). When used as a fermenter the main criterion for a good STR is that it maintains stable conditions, both in terms of the internal environment and in its ability to prevent microbial contamination. According to its mode of operation, an immobilized cell reactor may or may not be subject to infection with extraneous micro-organisms. Obviously, if sterility is to be maintained the immobilized cell preparation itself must either be sterile (for non-viable cells) or contain only the target immobilized cell species. Operation is always easier if there is no need to have the reactor, its sensors and connecting pipe-work sterile. Although sterilization is normally carried out with pressurized steam, other methods are possible. Krouwel, Van Der Laan & Kossen (1981), for example, briefly describe the use of 50% ethanol to sterilize immobilized spores of *Clostridium butylicum* and so avoid contamination during start-up.

In aerobic STR fermenters mixing is partly due to air bubbles moving up from the sparger outlets. The same effect can be obtained in an immobilized cell reactor but, because rapid growth is not required, the oxygen demand of the cells will be low. Low aeration rates will reduce problems of foaming. Low oxygen tensions may even stimulate product excretion and give positive benefits, as found by Vollbrecht (1980) for several species of aerobic bacteria. Linked to the low growth rate in immobilized cell reactors, there will be less metabolic production of heat during biocatalytic activity and this may either increase the need for heating (which is generally not of great economic significance) or reduce the need for cooling, which might be of benefit. Because most immobilized preparations are selected for use at as high a temperature of operation as possible the need for cooling is, in any case, less likely to arise.

In an ideal, completely mixed reactor there will be no diffusion limitation of biocatalytic activity because the concentration of the substrate will be constant throughout the reactor. With immobilized cells this can never be fully achieved and the reactor must be considered as heterogeneous, with a free fluid phase and a solid phase which may take the form of particles or films. Because the substrate is converted to the product within the immobilized cell preparation, gradients will exist from the completely mixed free fluid phase up to the sites of the biocatalytic reactions. With very small particles (of the order of macromolecular size) it is possible to achieve a constant supply of unchanged substrate and remove the products of the reactions at technically achievable flow rates of the fluid phase. With particles or films of sufficient size (0.1–1 mm diameter or thickness) to be of practical use in immobilized cell reactors, there is certain to be a zone of reducing substrate concentration as the surface is approached and into the centre of porous or gel support materials. With increasing fluid velocity the concentration adjacent to the surface approaches that of the bulk liquid. There is an upper limit, however, to the velocity of the flow of the free liquid phase which can be used. This is not generally due to difficulties in obtaining high liquid velocity but rather to the abrasion damage caused to the immobilized cell preparation. High stirring rates necessary to obtain completely mixed characteristics generate high shear forces at the surface of the impeller. This can be described by means of the Reynolds number (N_{Re}) which is a dimensionless group according to the equation:

$$N_{Re} = V d_i / \nu_k$$

where V = speed of impeller tip, d_i = impeller diameter, ν_k = kinematic viscosity of liquid phase.

From this it is obvious that the motion of the impeller is the most important influence (Figure 7.5). The actual shear velocity V_s is described by:

$$V_s = V/x$$

where V = speed of impeller tip, x = distance between impeller tip and vessel wall.

Klein & Eng (1979) describe a generally applicable method for the determination of abrasion. Experimental determination can be carried out in a laboratory-scale reactor or the actual working vessel. In all cases it is important to specify such items as impeller size and design, rotational speed, dimensions of the vessel, particled size and form, amount of particles present, the amount and nature of the liquid phase present and the duration of the experiment, which should obviously be carried out at constant temperature. The presence of gas bubbles will influence the

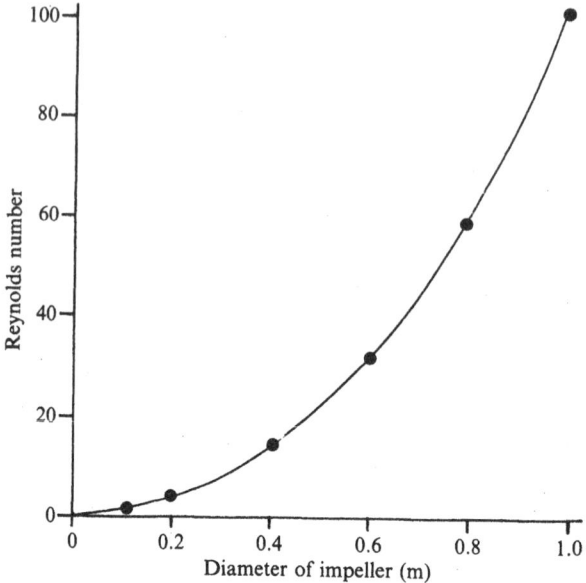

Figure 7.5. Graphical representation of the influence of impeller diameter on the Reynolds number of a stirred-tank reactor running at 100 rpm.

results and most determinations are carried out without aeration. An external loop, through a recording, broad-band, visible-wavelength spectrophotometer (or similar system) can be used if the abraded particles are in suspension. Using *E. coli*, immobilized in epoxy-resin beads, it was found that above a certain value for the cell loading in the beads, a dramatic increase in the abrasion occurred. Likewise, there was evidence of a critical value for the stirrer speed, although in general abrasion increased with increasing tip speed. Particles of 1 mm diameter or below were much less abraded than those of greater size. Increasing the number of particles present per unit working volume of the vessel also gave more abrasion.

Where attempts are made to scale up the results of laboratory experiments in STR systems, certain important features have to be considered, as discussed by Einsele (1978). For geometrically similar STR vessels it was found that power input per unit volume decreased with increasing volume while the Reynolds number increased, as did the mixing time (Figure 7.6). The data were collected from actual industrial bioreactors (fermenters) and showed also that impeller tip speed was relatively independent of volume. It was pointed out that, in practice, most large vessels are not based on predicted scale-up from geometrically similar small ones. Often the research involved the reverse of scaling-down to allow laboratory experiments under similar conditions.

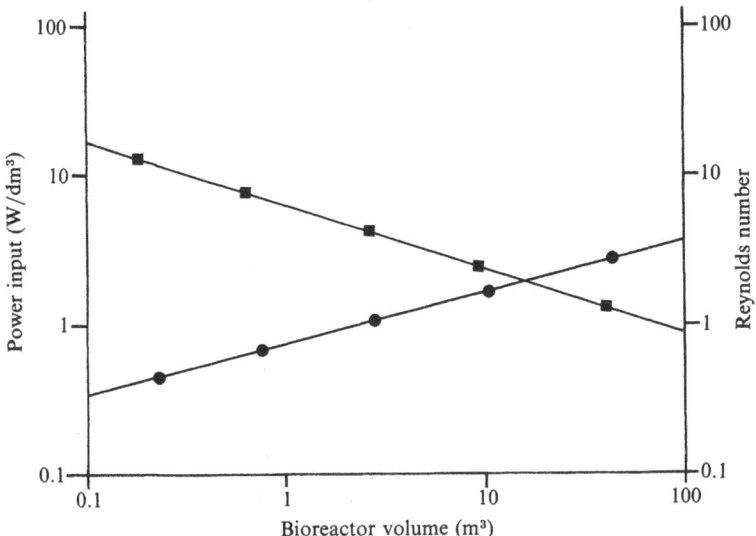

Figure 7.6. Graphical representation of the relationship between reactor size, power input (■) and Reynolds number (●) for actual industrial bioreactors (after Einsele, 1978).

The major importance of biological factors in scale-up calculations was emphasized.

Using a conventional one-litre magnetically stirred STR fermenter, Margaritis & Bajpai (1981) were able to obtain ethanol volumetric productivities of 13.5 g/l/h. The same calcium alginate-immobilized *Kluyveromyces fragilis* beads were used for eleven batch cycles with negligible reduction in the ethanol yield (96% of theoretical), although the rate of ethanol formation decreased, growth was obviously occurring and cell leakage became progressively greater. Compared to ordinary free cell fermentations there were benefits in the reduction of batch time, high production rates, repeated use of the cells and easier recovery due to the low cell concentration in the drained liquid medium. Unfortunately, the details of the impeller design and mixing regime during batch operation are not given, but the paper does illustrate that even relatively fragile immobilized cell preparations can be used in an STR. The beads were rehardened in 4% calcium chloride solution. Even higher ethanol productivities, in the region of 80 g/l/h, were obtained, using a flocculent form of *Zymomonas mobilis* in a continuous STR with mild agitation and gassing with nitrogen, by Fein *et al.* (1983). Biomass retention was by an overflow weir. This enabled dilution rates of 1.0 and above to be used, with glucose feeds of 100 g/l. This paper contains an interesting comparison between several different approaches to ethanol production using *Z. mobilis*, reworking the original data where necessary to obtain fair comparisons. As the authors pointed out, the use of naturally flocculating strains reduces costs by avoiding the need for cell recycle systems or separate immobilization stages. Looking to possible scale-up the CSRT system would have a lower nitrogen demand and probably more complete mixing than a gas-lift tower reactor using the same strain of *Z. mobilis*.

A more thoroughgoing analysis of the use of a CSTR for immobilized cells was presented by Krouwel, Groot & Kossen (1983). They studied the isopropanol–butanol–ethanol production by calcium alginate-immobilized *Clostridium beijerinckii* in a two-litre fermenter containing only about 400 ml of liquid and mixed by the impeller. Nitrogen gassing was applied to the medium reservoir and the atmosphere over the liquid in the reactor itself. Experimental results were used to test a mathematical model under continuous operation. The research represented a development from previous work using a column fermenter in which problems of incomplete mixing and non-steady-state conditions arose. Growth according to the Monod equation, the loss of new biomass from the beads to the liquid phase, and a steady state with no external mass transfer limitation were also assumed in the model. A marked effect of the start-up conditions was found, which was carried through during later operation and it appeared that the thickness of the active biomass region of the beads did not increase during continuous operation, presumably due to limitation of a critical growth component. Continuous rather than batch operation gave considerably higher volumetric productivity. The authors considered that better fit of the model to the experimental data required further knowledge of the properties of the bacteria, but that the prospects for this type of process were good.

A detailed kinetic study of β-glucosidase associated with polyacrylamide-immobilized *Alcaligenes faecalis* was carried out by Wheatley & Phillips (1983a), using a laboratory-scale batch STR. Particles of $0.25 \, \text{mm} \times 0.25 \, \text{mm} \times 2 \, \text{mm}$ showed negligible effect on the $K_m \text{app}$ as compared to free enzyme. Particles of $2 \, \text{mm} \times 2 \, \text{mm} \times 2 \, \text{mm}$ showed significantly higher $K_m \text{app}$ and the effectiveness factor was reduced to 0.71 at high and 0.27 at low substrate concentrations, indicating severe intraparticle diffusion limitation. Comparison of the effect of stirrer speeds of 200 and 800 rpm for particles of $0.65 \, \text{mm} \times 0.65 \, \text{mm} \times 2 \, \text{mm}$ showed only small increases in the $K_m \text{app}$ for the slower speed, indicating minor contributions of external mass transfer. These results confirm what would be expected from theoretical predictions. The authors also compared their results with the performance of a 1 cm diameter glass column continuous flow system using the $0.65 \, \text{mm} \times 0.65 \, \text{mm} \times 2 \, \text{mm}$ particles. As expected, at low substrate concentrations the column showed a lower percentage conversion but at high substrate concentration there was negligible difference. With increasing flow rates in the column $K_m$app tended towards a constant value, but one higher than that of the free enzyme. In this research the kinetics were, in part, based on extrapolation and initial rates, so there is a marked contrast in the approach to that used by Krouwel, Groot & Kossen (1983). No growth occurred (in any case the viability of the cells with polyacrylamide immobilization is generally lower) and long-term stability, including the effects of abrasion, was not investigated. The results confirm the predicted behaviour under the conditions used but obviously a practical application of the immobilized cell preparation would display other features.

Fluidized-bed reactors

Gas-lift fermenters have rapidly developed and are now used for commercial processes. The major advantages claimed for these are the lower running and capital costs, due to the absence of a stirrer motor, lower chances of contamination due to the absence of a drive shaft port, lower shear stress and improved aeration characteristics, particularly with large height to diameter ratios.

Kloosterman & Lilly (1986) used an external loop, air-lift reactor, for the biotransformation of hydrocortisone by alginate-immobilized *Arthrobacter simplex*. As might be expected, the stability of the beads was considerably greater in the air-lift reactor than in a stirred-tank reactor. Over a specified range the mean circulation time of the beads was approximately inversely proportional to air-flow rate but above a superficial gas velocity of 0.38 cm/second a steady-state value was reached. The actual value depended on the liquid to bead ratio. From other work cited in this paper it was already known that above dissolved oxygen values of 25–30% of full air saturation the bioconversion is limited by steroid diffusion rates. Systems of this type have not been used often in work with immobilized cells but are perhaps worthy of greater consideration.

Because the oxygen demand of immobilized cells is generally low (or non-existent in the case of non-viable cells carrying out simple biocatalytic reactions) the gas-lift reactor has not been used much for immobilized cells. To achieve complete mixing, gas and liquid flow rates will still be sufficiently high to cause shear stress. The need to have immobilized cell preparations of appropriate density also limits the range of support materials that can be used. The fluidized-bed reactors, where liquid flow upwards promotes fluidization have, however, been used extensively with immobilized cells. They provide conditions which can be varied between a completely mixed STR and a theoretical plug-flow packed-bed column, and the precise conditions in the reactor can be modified within a wide range between these two extremes. As the flow rate of the liquid increases for a given type of particle, a situation is eventually reached where the pressure reduction through the mass of particles equals the weight per unit area of the bed, and the particles are suspended and no longer in continuous direct contact with one another. The stage just before this, where a high proportion of contact time occurs, gives an expanded-bed reactor. This state is most likely to occur with relatively dense support materials and low flow rates. At the opposite end of the spectrum will be light support materials such as hydrophilic gels used with high flow rates. This may lead to loss of the immobilized cell preparation by elutriation (wash-out) from the top of the vessel. This may be prevented by mesh-protection of the outlet or the provision of overflow weirs and expanded upper regions to the column where flow rate and hence suspension capacity are considerably reduced. In some cases the whole reactor takes the form of an inverted cone. Where the particles in suspension are of different sizes or densities it is obvious that stratification may occur, with the larger or denser particles towards the base. Where living cells are immobilized and growth takes place any cells which become detached will tend to rise in the fluidized bed and may be washed out completely from the reactor, even though the immobilized cell preparation is retained. Where the immobilized preparation can be deformed, by its own weight in a column, the fluidized bed will help to prevent pressure drops due to compaction.

From what has been said in the previous paragraph it is clear that one fluidized bed reactor may be very different from another. Petersen (1965) placed two-phase fluidized-bed reactors into four idealized categories, depending on the completely mixed or plug-flow characteristics of the solid and fluid phases taken separately. If aeration or internal gas production are included the resulting third phase adds another order of complexity. If fluidization is maintained by liquid recycle rather than straight-through flow, substrate and product distributions in the reactor may show additional effects. It follows from this that no one mathematical description can cover all systems, but that sets of equations between idealized extreme situations can be developed and appropriate ones fitted to experimental data.

At low concentrations Atkinson (1974) gives equations of the following form for fermenter performance at low concentrations:

$$C_o/C_i = \exp\left[-YA_sZ/\bar{u}\right]$$

where

$$Y = \left[(1/K_1) + (1/h)\right]^{-1}$$

C_i = inlet substrate concentration, C_o = outlet substrate concentration, A_s = area of support surface per unit liquid volume, Z = fermenter length, K_1 is a first-order rate coefficient, which varies depending on microbial film thickness, $\bar{u}$ = average fluid velocity, h = mass transfer coefficient; while at high concentrations:

$$Q(C_i - C_o) = [K_1 L/K_2]A_s Z$$

where Q = volumetric flow rate per unit cross-section area, L = wet microbial film thickness or half-length of entrapped cell particles, K_2 = a further biological rate coefficient.

The mass transfer coefficient (h) is very sensitive to the flow pattern and must be determined by an experimental method rather than by calculation from theory. This can be achieved by using support particles of the same type and size as those for the immobilized cells. These are loaded in a comparable way with a substance which is sparingly soluble and easily quantified in the liquid phase. As would be expected, increasing the agitation reduces the difference between outlet and inlet concentrations, while increasing the immobilized cell preparation's activity increases the difference, for low inlet concentrations. It is also obvious that to maintain a concentration gradient of substrate and product between inlet and outlet the liquid movement should be as close to an ideal plug-flow as possible, which will generally mean relatively slow flow rates. This also leads to differential local reactor environments, substrate-rich towards the inlet and product-rich towards the outlet. Where immobilized cell preparations are based on flocculation or gels which have little density difference from the fluid, it is obvious that particles must be of moderate size (say, minimum of 1 mm diameter) to allow reasonable flow rates at conditions approaching liquid plug-flow. Where plug-flow is approached the substrate concentration throughout the reactor will strongly reflect the inlet concentrations and it may be realistic to use a zero-order kinetic for the reaction. The relationship between flow rate and particle hold-up follows the equation of Richardson & Zaki (1954):

$$V/V_t = (1 - M/P_d)^n$$

where V = fluid velocity, V_t = terminal fluid velocity, M = immobilized cell preparation local concentration, P_d = density of immobilized cell preparation and exponent n depends on the Reynolds number. With appropriate values substituted it has been found that this type of equation can be developed to reasonably represent alcohol production by yeast in a fluidized (tower) reactor at various flow rates (see Atkinson, 1974). The production of alcohol and wastewater treatment are two of the major areas of application of fluidized beds but represent markedly different systems and approaches.

In the research of Vijaikishore & Karanth (1986) the air supply was used to suspend K-carrageenan-immobilized *Pichia farinosa* in a continuous fluidized-bed reactor. The target product was the aerobic formation of glycerol, but the substrate feed rate was not sufficiently high to fluidize the beads. Wash-out was prevented by physical restraint with wire mesh screens. There was, however, a clear indication of diffusional limitation in the supply of oxygen to the cells in the rather large (4–5 mm diameter) beads used.

Andrews & Przezdziecki (1986) developed effectiveness factors for microbial flocs, solid spherical supports, porous supports and supports, such as activated carbon, which can adsorb large quantities of substrates or products. Their calculations were related particularly to the design of fluidized bed reactors and they emphasize the interrelationship between the type of immobilization system and the reactor's design. The two obviously need consideration together rather than to attempt to optimize either component in isolation.

The use of inverted conical (or tapered) reactors adds an extra dimension to the complexity of modelling fluidized beds. The general principles and advantages have been discussed by Scott, Hancher & Shumate (1978). This design, with an expansion angle of a few degrees from the inlet, gives good flow-patterns with minimal turbulent mixing. Even small or low density particles can be retained at high inlet velocities because the fluid velocity decreases with increasing height up the reactor. This can result in improved performance of immobilized cell reactors by allowing greater operational bed expansion.

The use of fluidized beds with self-immobilized mixed microbial populations on dense support materials is an economically attractive proposition. Cheap supports such as sand, the absence of any requirement for sterilization and the high volumes of relatively low concentration wastewaters have led to the development of this type of reactor for domestic and industrial effluents. Cooper (1981) described the basic principles of what are called 'biological fluidized beds'. These are frequently based on sand particles and may operate under aerobic or anaerobic conditions with, or less commonly without, liquid recycle loops. Stephenson & Murphy (1980) investigated the kinetics of waste-water denitrification using a column fluidized bed at pilot plant scale (0.14 m × 3.66 or 4.88 m). The support material was initially granular activated carbon but this was found unsatisfactory after three months of operation and replaced by sand. The effective particle sizes were 0.55–0.65 mm and 0.75–0.85 mm respectively and the wetted density 1.3–1.4 and 2.65 g/cm^3 respectively. The bottom of the column was fitted with a perforated plate to achieve good liquid flow distribution. The self-selected microbial population grew as a film on the support particles and the film thickness was kept within the range 0.10–0.15 mm by a four-bladed impeller at the top of the column. This sheared off biomass by abrasion and also allowed entrained gas bubbles to escape. The input was a completely nitrified effluent from an extended-aeration activated sludge treatment plant, although during kinetic experiments potassium nitrate was added. The work carried out was particularly interesting for its attention to the effect of temperature in the range 4–27 °C. Despite the perforated plate a distinct conical region (1 m high) of greater turbulence was visually obvious at the inlet, but tracer experiments indicated no short-circuiting or dead space and an idealized plug-flow model could be used for the overall kinetic study. It was found that a half-order kinetic model with an Arrhenius temperature effect was satisfactory to describe denitrification. Neither hydraulic flux in the region 0.25–1.7 $m^3/m^2/min$ nor type of support was significant in controlling rates of denitrification. The performance of the fluidized bed system was considerably better than that for either suspended growth or rotating biological contactor systems.

In a study of anaerobic treatment of a simulated wastewater Bull, Sterritt & Lester (1983) used a laboratory-scale (5 cm × 200 cm) fluidized-bed reactor. The support material was sand (0.22 mm mean diameter) and the column was filled to approximately 1 m settled height and run at approximately 20% bed expansion. The self-immobilizing inoculum was anaerobic digester sludge. Four start-up regimes were compared. These were a continuous loading of 4.6 kg $COD/m^3/day$ at 1.25 g/dm^3 reducing to zero at 35 days and stepped loading from 0.86 to 4.6 kg $COD/m^3/day$ over 35 days, with or without the same methanol additions as the continuous regimes. As might be expected, it was found that the stepped loading with methanol addition gave the most satisfactory start-up and consistent COD reduction over the initial period, although all regimes tended towards a similar performance at around 40 days' operation.

The process of nitrification was kinetically modelled by Denac *et al.* (1983), testing the model with data from a fluidized-bed reactor with sand as the biofilm support material and a simulated wastewater containing ammonia. Diffusion limitation was found to be due to either oxygen or ammonia, depending on their relative concentrations, but for each situation half-order reaction kinetics were sufficient to describe the apparent overall rate of nitrification. Depending on the biofilm area, oxygen concentration and flow rate, significant quantities of nitrite were present in the outlet stream. A four-component, diffusion-reaction model with double Michaelis–Menten kinetics was used.

Royer *et al.* (1983) used alginate-immobilized beads of *Coriolus versicolor* in a specialized fluid and bubble-lift fluidized bed to decolorize Kraft paper-mill effluent, with sucrose (50 mM) as an added carbon source. The reactor consisted of four linked columns with the first fed by a peristaltic pump and the others by a stepped gravity overflow configuration. Air was released from a pinhole sparger close to the base of each column to provide a dissolved oxygen concentration of greater than 4 mg/dm³. The same amount of beads was used in each column. Assuming negligible resistance to internal diffusion in the pellets and radially, uniform concentration and constant microbial activity, the reaction was found to be approximately second order (1.85). Calculated values of decolorization gave reasonably good fit to the experimental data using the approximate mass balance calculations.

The production of solvent or fuel alcohol in fluidized-bed reactors can be roughly divided into those processes based upon yeasts (mainly *Saccharomyces* species) and those using bacteria (particularly *Zymomonas mobilis*). Prince & Barford (1982a) used a glass laboratory-scale tower reactor of 7.5 cm diameter with a 15 cm diameter expanded upper settling region with a baffle and an overall height to diameter ratio of 22:1. Naturally flocculent *Saccharomyces diastaticus* and *S. cerevisiae* were used and high (70–90 g dry wt/dm³) cell densities could be achieved. Adequate aeration was given for start-up growth and reduced to minimal for the ethanol production phase. At each glucose feed concentration tested (121, 152, 199 g/dm³) the flow rate was increased until washout or significant non-conversion of substrate occurred. The latter was of greater potential commercial significance. Maintenance of aseptic conditions for up to 14 weeks was not a problem and mild bacterial contamination could be removed by increasing the flow rate and lowering the pH value to 3.5. The yeasts used were selected for good flocculation rather than high ethanol production, but using a similar reactor the same authors (Prince & Barford, 1982b) were able to demonstrate a general method for selecting flocculent forms from non-flocculent strain of *S. cerevisiae*. Initial start-up (72 hours) was at a dilution rate of 0.06/h during which time the reactor was operated as a well-mixed vessel. The tiny flocs which then began to form were encouraged by step-wise increments of dilution rate up to a maximum of 0.27/h when 118 g dry wt cells/dm³ was obtained, in flocculent form. Chen & Zall (1982) used a tapered (inverted cone) expanded-bed reactor to study the continuous conversion of whey into alcohol by self-immobilized *S. fragi*. Cellulose acetate powder was used as the support but it was suggested that a coarser grade would have helped prevent wash-out, which was aggravated by CO_2 production by the yeast. A productivity of approximately 7 g ethanol/dm³/h was obtained. Comparison was made with a two-part membrane fermenter system in which a simple ultrafiltration tube (MW cut-off 10 0000) was used as part of an external recycle loop, with the culture passing upwards through the tube lumen. In this case, however, reagent grade lactose was used in place of whey solids but a similar productivity was achieved and the authors concluded that both systems would deserve further study.

A comparison of three laboratory-scale fluidized-bed reactors for continuous ethanol production by a flocculent strain of *Z. mobilis* was made by Strandberg, Donaldson & Arcuri (1982). Particular attention was paid to alternative configurations allowing simple degassing of the liquid. These were a straight vertical column, a vertical tapered (inverted) column with an upwards-angled side-arm acting as a gas-free settling region and a 45° slope multi-vented column. In the reactors, initial wall film growth detached and became flocs (approximately 1 mm diameter). The angled side-arm proved satisfactory in preventing floc wash-out. The sloping column did not give good mixing or separation of the flocs from the effluent. In the experiments productivities were not optimized, although some provisional calculations were made showing 0–64 g ethanol/dm^3/h for the data quoted. Although the side arm system is convenient in glass laboratory-scale reactors there would appear to be constructional advantages in using the expanded upper diameter section, with internal settling baffle and overflow weir, for larger-scale systems as in the work of Prince & Barford (1982c) with *Z. mobilis*. In the latter work productivities of 100 g ethanol/dm^3/h were obtained and the limitation to performance was incomplete conversion of the glucose feed rather than wash-out at high flow rates.

Vermiculite of 250–380 μm size was chosen by Bland *et al.* (1982) as a self-immobilization support for *Z. mobilis* to use in a continuous inverted cone, fluidized-bed reactor. This had a simple overflow weir and liquid recycle. The recycle was used to maintain bed expansion at low reactor feed rates. The recycle passed through a simple clarifying unit which assisted complete degassing and allowed recovery to the recycle stream of any immobilized cells washed out of the reactor. Productivity based on reactor volume was a maximum of 105 g ethanol/dm^3/h at a dilution rate of 3.6/h, although glucose conversion was only 64.2%. It was suggested that a higher productivity might be achievable. Such a low conversion would not be commercially acceptable. Klein & Kressdorf (1983) used a three-stage inverted cone fluidized-bed reactor system for *Z. mobilis* immobilized in calcium alginate beads. Due to gas production and associated turbulence the first two vessels had continuous flow properties. This gave a higher productivity at high percentage conversion of the glucose than six previously reported systems with which it was compared. As was pointed out by these authors the basis of claims for high productivity needs careful scrutiny because many experiments are unrealistic with regard to sugar conversion at high feed concentrations and flow rates.

Krouwel, Van Der Laan & Kossen (1980) used a simple inverted, slightly tapered laboratory-scale column to study the feasibility of *n*-butanol and ispropanol production by growing calcium alginate-immobilized *Clostridium butylicum*. The shape was selected to minimize gas slug formation, although neither this aspect nor maximum productivity were optimized in this work. Leakage of biomass did occur, but the effluent cell concentration was less than half that found in a batch fermentation. Two sizes of cylindrical column, with sintered glass air inlet plates across the bottom, were used by Kennedy *et al.* (1980) to study wort (ethanol) conversion to vinegar (acetic acid). They used two strains of an *Acetobacter* species, one producing polysaccharide and described as aggregating and the other a non-producer of polysaccharide. It was found that biomass retention of the former strain was improved by the addition of hydrous titanium (IV) oxides but that for the latter strain a titanium–cellulose powder chelate was needed. This led to higher acetic acid productivities and the possibility of using higher medium throughputs.

Margaritis & Wallace (1982) used a fluidized-bed reactor specially designed to ensure complete mixing in their work on *Z. mobilis*. Unlike conventional air-lift fermenters this consisted of two concentric draught tubes inside the reactor, with four jets of incoming recycled medium released between the inner and outer draught tubes. The *Z. mobilis* were immobilized in calcium alginate beads. The system was compared with a similar immobilized cell preparation in a magnetically stirred reactor of conventional design. As expected considerable leakage of biomass occurred in the STR system (at 300 rpm) but little leakage occurred in the fluidized-bed system. The latter showed a higher ethanol production rate. Repeated batch operation of the fluidized-bed system with the same beads was also possible. Biomass leakage increased slowly with repeated batch reuse. The type of low shear, completely mixed, fluidized-bed reactor used here could be of interest in the production of secondary products and has the advantage of being carried out in a vessel which can be relatively easily converted from a conventional STR. In general, alginate is both too expensive and too fragile an immobilization support for low-value products such as ethanol.

Klein & Kressdorf (1986) described a three-stage reactor system for ethanol production, with two tubular reactors, allowing good CO_2 removal followed by a plug-flow reactor. *Z. mobilis* was immobilized in alginate or K-carrageenan. This system made maximum use of the operational characteristics of the different types of reactor.

Atkinson *et al.* (1980) gave a good account of the evolution and use of general-purpose, weft-knitted stainless steel mesh systems in a fluidized-bed, completely mixed microbial film fermenter.

A particular type of reactor which provides good mixing of heavy immobilization particles, such as spheres of knitted stainless steel, is the spouted bed reactor. Webb, Fukuda & Atkinson (1986) describe its use for cellulase production. A powerful basal jet of recycled liquid carries the particles upwards in the centre of the reactor, to be replaced by downward movement of particles at the periphery of the vessel. The abrasion of one particle against another causes any excess biomass developing outside the sphere to be sheared off, giving a constant microbial biomass once the spheres are fully colonized. This system has proved very useful for filamentous fungi such as the *Trichoderma viride* used in this example.

Rotating thin film reactors

A completely different approach to the problems of ensuring adequate conditions for the growth and metabolism of immobilized cells is to use a rotating surface of relatively large area instead of small discrete support materials. One system, discussed by Moser (1982) uses a tubular support surface while the other makes use of rotating discs placed on a central shaft (see, for example, Blain *et al.*, 1979). The rotational speed in both cases can be low and for the disc reactors is not normally greater than 10 rpm. The systems operate horizontally so that it is possible to adjust the height of the liquid in the reactors to obtain the desired amount of aeration (emergence) for every rotation. The disc concept has been extended by Chang, Joo & Ghim (1984) to packed discs. This allows the use of any form of artificially immobilized cells (their work was with purified enzymes in calcium alginate beads) in reactors of this design. A major feature of rotating film reactors is that they operate at very low shear rates. They can be run in either batch or continuous mode. Although their design and operation is simple and reportedly very reliable, they have not found extensive application. This probably reflects the fact that most are manufactured specifically for the user and are not readily available 'off the shelf', rather than due to any inherent operational problems.

Moser (1982) distinguished a selection of different types of rotating surface reactors with thin-layer characteristics. In the thin-layer film reactor (TLFR) the inner cylinder rotated in a trough of medium and became coated on its outer surface with microbial growth. In the horizontal rotary reactor (HRR) a single cylinder contained the microbial culture and the self-immobilized cells formed a film on the inner surface of the cylinder. The film of liquid is continuously renewed as the rotation takes place and the mass transfer coefficient at the gas–liquid interface is directly related to the rotational speed. With the double cylinder system (TLFR) there is an approximately constant shear rate across the gap between the two cylinders, provided the gap is small. Plug-flow characteristics could be achieved in continuous mode in the TLFR but not in the HRR, where bulk mixing occurs. With *Aspergillus niger*, for example, it was found that the film thickness could be controlled in the region up to 2 mm thick by the rotational speed. Above 50 cm/sec peripheral speed, the film was below 1 mm thick. This type of reactor is also suitable for viscous media which are a problem in conventional STRs. Aeration and gas production have much less effect on plug-flow than in other designs of reactor.

The constriction of a laboratory-scale disc reactor is described by Blain *et al.* (1979). The discs were made of high-density roughened polypropylene and some had removable segments which could be used for sampling biomass production. Several species of filamentous fungi were allowed to self-immobilize onto the discs. Contamination was not found to be a problem. Discs were rotated at 8 rpm and air supplied at 2 dm^3/min to the upper part of the reactor. After the initial rapid growth phase, during which the discs were colonized by the fungi, it was possible to maintain dissolved oxygen above 20% saturation by this simple aeration system. When *A. niger* was grown in a sucrose-rich medium (140 g/dm^3) film thicknesses of 5–10 mm developed. The films were heterogeneous but there was evidence of the inner layers still being viable even after 15 days' growth. Results of methylene blue uptake into and loss from the microbial film liquid phase indicated a substantial rate of interchange with the bulk fluid. Artavanis & Todd (1980) used the sulphite oxidation method to investigate oxygen transfer in a rotating disc reactor. They found the method unsuitable, giving peak uptake rates within the first 10 minutes which reduced to a relatively steady rate after two minutes. It was concluded that the sulphite method is not suitable for measuring true mass transfer coefficients in disc reactors due to enhancement by chemical reaction. By using batch replacement of medium deficient in nitrogen and phosphorus it was possible to restrict growth on the discs (Anderson *et al.*, 1980) to prevent excessive film thickness and obtain good yields of citric acid. The conditions affecting production could be easily studied by this same replacement technique, because the active biomass is retained when the reactor contents are removed.

In their simplest form, rotating disc reactors are particularly suitable for the growth of self-immobilizing filamentous micro-organisms but are less suitable for bacterial films which slough easily. Livesey-Goldblatt *et al.* (1977) compared four types of film reactor for the oxidation of acidified ferrous sulphate solution. The fastest specific oxidation rates (g/m^2/h) were obtained with a submerged corrugated plastic packing with forced aeration, followed respectively by plate and spray, rotating disc and trickle flow. The forced aeration system allowed higher dissolved oxygen concentrations to be maintained, which probably accounted for the success of the system for this particular application. A steady-state kinetic model was developed by Watanabe, Ishiguro & Nishidome (1980) for nitrification using a rotating disc reactor. Zero-order reaction kinetics were found in the biofilm but at low ammonia concentration this changed to first order because of rate-limiting transport processes. Experimental data supporting the model were obtained under a variety of operational conditions, using the reactors in a completely-mixed flow mode. In further experiments with *A. niger*, Anderson *et al.* (1981) used the rotating disc reactor to remove carbohydrate from a simulated wastewater. The increase in attached biomass with time was approximately linear over a wide range of conditions and glucose removal was virtually complete up to input rates of 1 g/dm^3/h. When excessive overgrowth occurred this could be stripped off and easily recovered but no growth occurred in the free bulk liquid of the reactor. When a mixed-culture sewage sludge inoculum was used, growth occurred throughout the reactor leading to continuous release of biomass in the effluent, giving the fungal system considerable advantage.

Membrane reactors
Ultrafiltration units have been extensively used as enzyme reactors and to a lesser extent as immobilized cell reactors. For the former application kinetic modelling has been carried out by several different workers. Flaschel & Wandrey (1979) discuss the general principles. To separate product from cells there must clearly be a driving force, in the form of either a concentration (diffusive) or a pressure (convective) gradient. For immobilized cell applications a typical system consists of a recycle vessel connected via a pump to the core space of the hollow fibres and the cells are in the outer shell compartment of the hollow-fibre unit. Roy *et al.* (1982) compared units with 60, 108 or 300 fibres, each of 100 μm internal diameters. The organism was *Lactobacillus delbrueckii*, inoculated into the outer shell. The unit has been previously sterilized with ethylene oxide and wetted with 50% (v/v) ethanol. Flow rates of the medium varied between 60 and 200 cm^3/h. Results were compared to a batch conventional fermentation for lactic acid production. Very high cell densities were obtained in the shell compartment of the hollow-fibre unit (480 g dry wt/dm^3) compared with a batch fermenter (9 g dry wt/dm^3). This resulted in very high volumetric productivities for the hollow-fibre unit although the specific productivity was much less than for the batch fermenter. It was concluded that the membrane restricted substrate supply (glucose as major carbon source) to the cells. Towards the end of the run the fibres were distorted by bacterial growth and penetration into the fibre cavity and recycle stream occurred after about 100 hours' operation. The rate of lactic production did not increase proportionally with increase in fibre surface area, as would be expected if membrane mass transfer was limiting.

Using a regulatory mutant of *E. coli* in a hollow fibre reactor, Paterson, Fane, Fell & Rogers (1986) found that even with pure oxygen aeration there was evidence of oxygen limitation. The fibres were isotropic, hydrophobic, polypropylene of 0.2 μm nominal pore size.

Pierrot, Fick & Engasser (1986) considered that an external hollow-fibre module, with cell recycle to an STR, was a promising system for the continuous production of solvent (acetone and butanol) by *Clostridium acetobutylicum*.

Engineering aspects of hollow-fibre reactors and modes of operation have been briefly described by Breslau & Kilcullen (1975), while Flaschel & Wandrey (1979) give more information on membrane properties and kinetic aspects. Cellulose acetate membranes are very susceptible to microbial damage, cannot be steam sterilized and are unsuitable for use outside of the pH range 4–9. Polyamide is more satisfactory at extreme pH values but cannot be steam sterilized. Polysulphone membranes are the most stable and can be steam sterilized. Although ethylene oxide could be used to sterilize all types of membrane its toxicity makes it hazardous to use in most laboratories. The use of aqueous formaldehyde is generally preferred, with thorough washing with sterile water after sterilization.

Using a hollow fibre reactor, Altshuler *et al.* (1986) were able to obtain a several-fold increase in the IgG volumetric productivity (260 μg/ml) of a murine hybridoma cell line compared with spinner flask culture. The system showed good IgG production over relatively long periods. Tharakan & Chau (1986b) discussed the problems of using hollow-fibre reactors for immobilized cells and carried out a detailed study of the pressure distribution. Cross-flow operation was suggested as a means of overcoming the problems of unequal substrate supply and consequent uneven growth and metabolism. Tharakan & Chau (1986a) describe the successful use of a radial (cross) flow hollow-fibre reactor for mammalian cells. Despite the work involved, some researchers have preferred to develop their own membrane reactors rather than accept the commercially available hollow-fibre devices. Wisniewski *et al.* (1983), for example, prepared their own polysulphone membrane *in situ* on the surface of the sintered polyvinylchloride support forming the entrance to a funnel-shaped insert into a laboratory-scale reactor. This had an inverted taper to the lower portion and was magnetically stirred. The organism in the reactor was *Rhodotorula mucilaginosa* carrying out the continuous transformation of benzaldehyde to benzyl alcohol. Gas input to the reactor headspace provided the motive pressure for continuous liquid flow across the membrane. The kinetics conformed to those of a continuous completely mixed reactor. The apparent V_{max} was directly related to the biomass concentration. The rejection coefficient for the cells was over 99% but some penetration occurred, with loss of active biomass during continuous operation. No growth substrate was supplied to the reactor, so biomass did not increase.

It is obviously possible to construct an enormous variety of membrane reactors in the laboratory. How many of these could be scaled up to economic production volumes is a matter for considerable debate. In general, other solutions to the problems of using immobilized cell reactors have been taken. The membranes which are currently available have been designed for ultrafiltration and are excellent for use in enzyme reactors. There will not be a single type of membrane suitable for all immobilized cell preparations nor for self-immobilizing all types of cells. Some general-purpose units could be designed, having hollow fibres with large pore diameters, good mechanical strength, ease of sterilization and resistance to microbial degradation and surface growth, which tends to block the pores. The outer shell compartment would benefit from specific attention to increasing its mixing characteristics. Such reactors are not yet available and their construction is probably beyond the manufacturing capabilities of most laboratories. When they do become available they will have excellent research potential.

Fixed-bed support systems

Column reactors. Cells immobilized onto particulate support materials can be packed into a column and liquid containing substrate passed through the column in an upward or downward direction. If the profile of the fluid velocity as it moves through the column is perfectly flat the result is perfect plug-flow. The volume of a column occupied by the immobilized cell preparation, in the packed region, will depend on the shape, size and size distribution of the particles and the evenness of packing. In general it would be expected that spherical particles of a single diameter would pack most evenly, provided the column diameter is considerably greater (say of the order of 10 times) than that of the individual beads. Size distribution can be assessed by direct observation or appropriate automatic instruments. The void volume of a column is the liquid phase existing between the particles and in an immobilized cell reactor no biocatalytic activity will occur in it, only mass liquid flow. Some of the space in macroporous or mesh particle supports will also be void volume. The void volume in the reactor is most easily measured by displacement of liquid on immersion, either *in situ* in the column or using a representative sample under similar conditions. Too high a void volume will obviously produce low conversion rates per column unit volume. Too low a void volume will restrict liquid flow.

The pressure drop from inlet to outlet in a column is an important operational parameter of a packed-bed column. In general, the pressure drop will be larger, the greater the height of the column. If the particles are compressible this will be an added difficulty. A pressure drop of approximately 21 kPa is acceptable for most columns, giving a maximum working height of approximately 4.6 m, with non-compressible spheres. Above this pressure drop there is a very rapid rise in flow resistance and operational problems are almost certain. Some commercial process columns may show pressure drops as high as 1 bar. The influence of compression will be time-dependent and in a down-flow column can at first be reversed by backwashing. Some support materials may, however, become irreversibly deformed. Upflow operation will keep a bed from compacting, but the flow rate must be kept at an appropriate value if expanded or fluidized-bed operation is not desired. Where inlet streams contain particulate matter, the immobilized cell preparation contains a proportion of small particles, or new cell growth occludes the space between particles, pressure drops are likely to become excessive. The theoretical and practical determination of pressure drop in both compressible and non-compressible beds has been described in detail by Buchholz & Godelmann (1979) and Klein & Kluge (1979). For non-compressible spheres and approximately for granular beads the pressure drop ΔP for laminar flow conditions is given by:

$$\Delta P = 160[\{(1 - 2V_v)^2/V_v^3\}\eta u(H/d_p^2)]$$

where V_v = void volume of column, η = viscosity, u = linear flow rate, H = bed height, d_p = particle average diameter.

Related equations can be derived for turbulent flow conditions. With compressible supports the pressure drop will increase with time and is also not a linear function of flow rate (Figure 7.7). Such experiments should therefore be carried out under the conditions of the intended use of the immobilized cell preparation. Buchholz & Godelmann (1979) suggest that considerable advantages could be gained by having an incompressible core around which compressible immobilization supports could be used.

The differential mass balance for a plug-flow reactor can be described by the equation (Atkinson, 1974):

$$\bar{u}(dC/dZ) = -R_v$$

where $\bar{u}$ = average fluid velocity, dC/dZ = concentration gradient of substrate in the direction of flow, R_v = reaction rate per unit fluid volume. If values are available for R_v the length of the reactor needed to give a specified productivity can be determined.

The effective rates of reaction will be markedly influenced, in most practical situations, by the external mass transfer of substrates and products in the bulk liquid and their intraparticle mass transfer. The latter is described as the microenvironment and the factors influencing it are combined in the dimensionless Thiele modulus. The former is called the macroenvironment and the factors influencing it are described by the

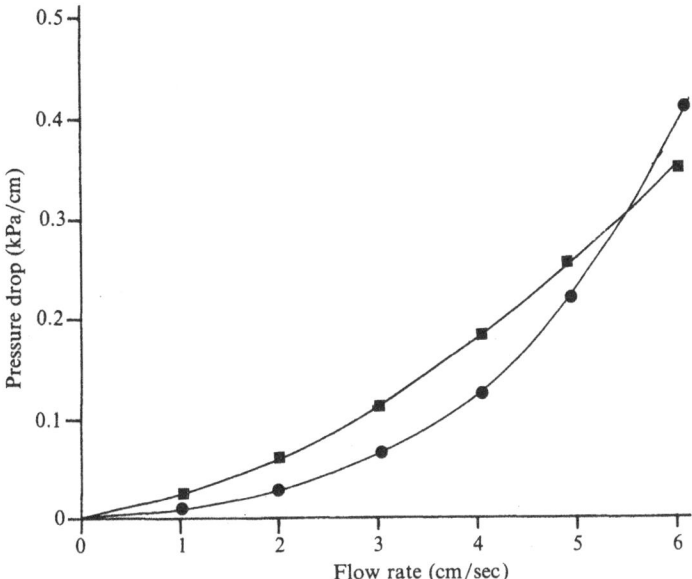

Figure 7.7. Graphical representation of the relationship between flow rate and pressure drop for a packed column containing (■) non-compressible beads and (●) compressible beads of calcium alginate. The former had been air-dried before use. The reactor was a packed bed (after Klein & Wagner in Buchholz, 1979).

Sherwood number Sh (another dimensionless modulus). This is given by Kasche & Buchholz (1979) as:

$$Sh = (K_s d_p)/D_s$$

where K_s = mass transfer coefficient, d_p = particle diameter, D_s = substrate diffusion coefficient.

The external mass transfer limitation has been discussed in detail by Buchholz (1979) and the internal diffusion limitation by Kasche (1979). The latter author draws a distinction between a stationary effectiveness factor for immobilized biocatalysts, which is simply the ratio between reaction rates of immobilized and free biocatalysts and the operational effectiveness factor, which is the ratio of the time required to convert a specified amount of substrate with the same amount of free and immobilized biocatalyst. This is determined experimentally and can be used to compare different types of reactor. The variation of Sherwood number with the relative velocity of the particle is shown in Figure 7.8.

The influence of internal and external diffusional limitations on the β-glucosidase activity of polyacrylamide-immobilized *Alcaligenes faecalis*

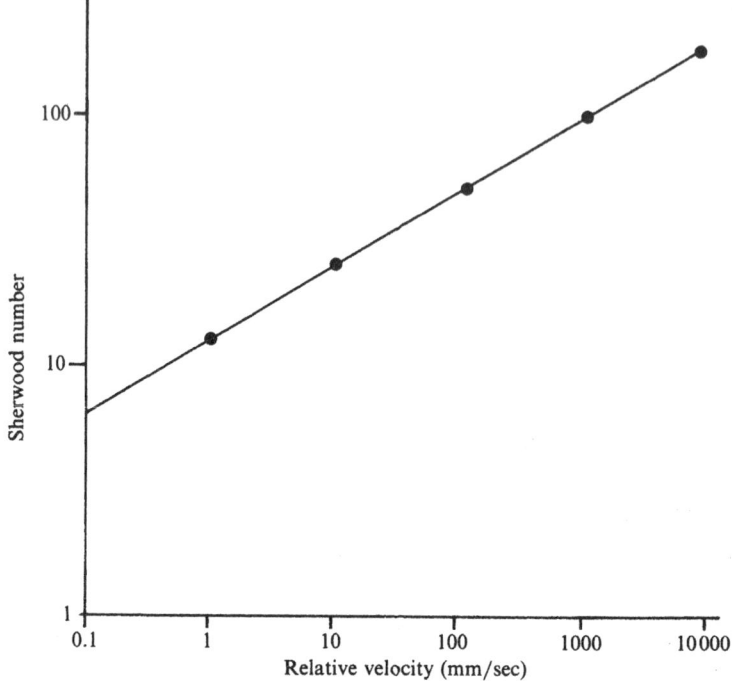

Figure 7.8. Graphical representation of the relationship between Sherwood number and the relative velocity of particles to bulk liquid flow, for particles of 0.5 mm diameter and a substrate diffusion coefficient of 5×10^{-6} cm^2/sec (after Buchholz, 1979).

was studied by Wheatley & Phillips (1983a). Comparison was made between a laboratory-scale stirred vessel and an upflow packed column. At high substrate concentrations there was no difference in percentage conversion for a given residence time, but at low substrate concentrations the column gave a lower conversion due to external diffusion limitation. Initial rates of reaction in the column were, as expected, related to the flow rate. The apparent K_m was constant for a given flow rate and decreased with increasing flow rate but tended towards a constant value considerably higher than for free enzyme, showing that internal diffusion limitation was also important. These results are typical of those to be expected in a fixed-bed column. Grote *et al.* (1980) found that channelling and compaction were problems in a fixed-bed column of alginate-immobilized *Z. mobilis*. They used perforated plates to divide the laboratory-scale reactor into sections and comparison was made with the same culture immobilized by K-carrageenan onto Raschig rings. The latter gave slightly higher ethanol concentrations and a volumetric productivity of 53 compared to 44 g ethanol/dm^3/h for the alginate system, a practical demonstration of the Buchholz & Godelmann (1979) concept of incompressible inner cones. Compared to free cells the specific rates of glucose uptake and ethanol production were reduced significantly. To overcome the same problems of compaction and channelling, McGhee *et al.* (1982) used a central shaft carrying a series of perforated discs inside a laboratory-scale glass column. *S. cerevisiae, S. uvarum* and *Z. mobilis* were immobilized in alginate. For the yeast cells the disc shaft resulted in an approximately three-fold increase in alcohol production but for the *Z. mobilis* there was no significant improvement.

A different approach to column operation was taken by Klein *et al.* (1979) who immobilized *Candida tropicalis* in a variety of different gels. Phenol degradation was the test biocatalytic reaction and a separate external aeration column was included in the recycle system. The immobilized cell column was operated at a high recirculation rate to achieve completely mixed reactor conditions. A thorough discussion of the kinetics of the system was included in the paper and it was found that the experimentally determined values for the efficiency factor were somewhat lower than those calculated on the basis of the Thiele modulus approach. An interesting separation of different phases of organic waste conversion to methane was described by Messing (1982). The hydrolytic-redox stage was carried out in a vertical, upflow column and the anaerobic methanogenic stage in a following horizontal column. Sewage micro-organisms were self-immobilized in controlled-pore extruded Cordierite of approximate dimensions 2 mm $\times$ 2–6 mm. This support was compared to similar-sized extruded brick containing carbon and larger (approximately 2 mm $\times$ 10 mm $\times$ 20–50 mm) segments of insulating fire-brick. The substrate was pre-filtered raw sewage with the pH value adjusted to 8.6–8.9. It was calculated that more than 90% of the substrate organic carbon not required to support bacterial growth was converted to methane and the percentage of methane in the evolved gas was also over 90%. The excellent properties of the laboratory-scale system were attributed to high cell densities and prevention of wash-out, plus the pre-adjustment of input pH value, moderate pressurizations and good methane removal in the horizontal column. The extruded brick support gave slightly better results. In a similar approach to the efficient removal of cell-generated gaseous products of metabolism, Margaritis *et al.* (1981) used an almost horizontal column with a central stainless steel fine mesh grid dividing it internally into two chambers. The liquid phase was 41% of the total reactor volume. This overcame channelling problems which arose when the column was used vertically. The organism studied was *Z. mobilis*, immobilized in alginate for ethanol production. Productivity was 102 g ethanol/dm^3/h for a 100 g glucose/dm^3 feed with 87% conversion. Intermittent treatment with calcium chloride solution to stabilize the beads gave productivities up to 116 g/dm^3/h. Open tubular reactors do not appear to have found application with immobilized cells but have been extensively used as enzyme reactors. The reader is referred to Pedersen & Horvath (1981) for a major article dealing with such reactors.

Down-flow stationary fixed-film reactors
If scale of operation is taken as the criterion, the down-flow fixed-film reactor is probably the most extensively used immobilized cell reactor system. It has the advantages of simplicity and cheapness and is used mainly for the treatment of low-value, large-volume wastewaters. The supports available vary from very cheap materials, such as coke, to numerous types and forms of plastics. Although generally operated under conditions which enable the self-selection of organisms from natural mixed populations, there is no theoretical reason why an appropriately designed system should not be capable of sterilization and used with single-species culture. Aeration is normally dependent upon shape, size and form of the support material, the thickness of the microbial film developed and the volume of water flowing through the system. Gravity down-flow is usual, but pressurized sprays may be used, with a consequent improvement in aeration as a result of transient aerosol formation. Counter-flow aeration from jets at the base of the reactor is also possible.

During operation biofilm development may become excessive and fill voids, resulting in decreased permeability and flow channelling. Although the principles of mass transfer, external and internal diffusion resistances, productivity evaluation and so on are similar to those of other reactors, the flow characteristics and substrate concentration profiles are more complex than those of completely mixed or plug-flow reactors. Atkinson (1974) provides an interesting discussion of the design principles of a trickle-flow reactor (fermenter), together with illustrations and some basic features of a range of common packing materials for these systems. Some of these supports are also used in column reactors. Packings may be of randomly arranged, relatively large (of the order of 3–5 cm) discrete particles or of regular, prefabricated sections with regular channels and high porosity.

Because the self-immobilization is on the surface of the support material, the specific area of the support is clearly important. Growth is unlikely on dry surfaces and so the mass flow of liquid per unit cross-sectional area of the support must be sufficient to wet the entire surface if maximal biomass is to be achieved. The value of the surface tension of the support and the viscosity of the liquid will have an effect. The development of excessive thickness of biomass can be prevented by hydraulic or mechanical scouring or by microbial interaction in the film. As the liquid passes over the surface of the biofilm its flow characteristics are generally considered laminar rather than turbulent, and the flow itself is planar over the outer microbial surface rather than partially submerged. The passage of the liquid from one support particle to another may cause turbulence, but this can often be neglected for the purposes of design calculations. Under this type of flow condition, the profile of substrate concentration passing downwards over the support will be of a pseudo-plug-flow type and a concentration gradient profile will also exist from the bulk flowing liquid through the microbial film to its basal attachment to the support. The fluid velocity profile u is expressed as:

$$u = u_s[1 - (y/\delta)^2]$$

where u_s = surface velocity of the film, y = the transverse space coordinate, δ = the liquid film thickness. The transverse flux of substrate (N_y) under these flow conditions would be due only to diffusion and could be described by Fick's Law:

$$N_y = -D(\partial C/\partial y)$$

where D = liquid phase diffusion coefficient, C = the local concentration, while the axial flow of substrate (N_z) is dependent only on the flow of liquid:

$$N_z = uC$$

where u = fluid velocity. C = the local substrate concentration .

This approach is extended by Atkinson (1974) to differential mass balance and design equations. The model was then tested with experimental data from a reactor containing 5 cm diameter wooden spheres. Hydraulic washing was used to maintain the mixed-culture microbial film to a relatively constant thickness. Consistent and reproducible data were obtained. The rate of substrate (glucose) removal tended to an asymptotic value with increasing inlet concentration and the efficiency of substrate removal accordingly fell at the same time (Figure 7.9).

A needle-punched polyester support was used by Duff & Kennedy (1983) in a down-flow fixed film reactor. Various recirculation ratios (from 0 to 500) were applied to the reactors for the removal of COD from a synthetic sugar waste, the loading rate of which was progressively increased during experiments. The recirculation ratio had no significant effect upon start-up loading rate increase or the efficiency of COD removal but a higher ratio generally increased the upper limit of loading rate which could be tolerated by the reactor. The flow was calculated to be laminar and, particularly at high recirculation ratios, the turbulence produced by gas production was insignificant. At high recirculation ratios it was concluded that mixing was considerably enhanced but the major

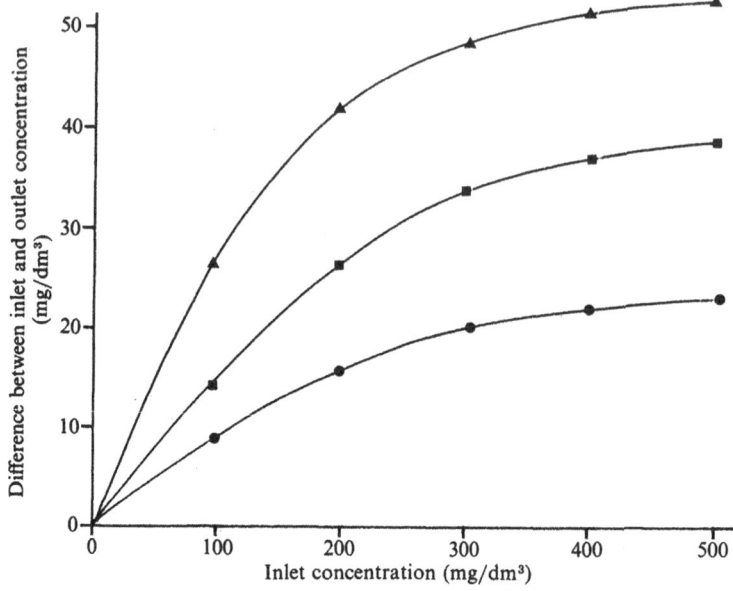

Figure 7.9. Graphical representation of the relationship between inlet concentration and substrate removal for a down-flow film reactor at flows of (●) 0.054; (■) 0.027; (▲) 0.015 cm³/cm²/sec (after Atkinson, 1974).

effect was probably due to more even distribution of the substrate over the support, giving more uniform development of the microbial film. Because the response to increased recirculation ratios was rapid, it was suggested that there was a reduction in internal diffusion limitation in the film. In previous work using the same reactor system, Duff & Kennedy (1982) had shown that when operated with a thermophilic mixed culture the reactor was less stable to hydraulic or organic overloading. It was suggested that this might be due to less diversity of species in the thermophilic mixed culture or their inability to cope with environmental stress. Conventional Arrhenius plots are of doubtful value in experiments with dynamically changing mixed cultures, unlike the situation with pure cultures.

8 Special systems

The major part of this book has been devoted to methods of immobilization and those applications in which the immobilized cells are used as biocatalysts in the production or biotransformation of substances. In this final chapter attention will focus on the use of immobilized cells to achieve other objectives. There is already an extensive literature on the use of immobilized enzymes as an aid to analysis. By making use of the specificity of enzyme catalysis it is possible to measure the concentrations of appropriate substrates (or a related group of substrates if the specificity of the enzyme is broad) in complex mixtures containing other substances. If a coenzyme-dependent enzyme is used it is possible to measure the concentration of the appropriate coenzyme by making its concentration the limiting factor controlling reaction rate. The increased stability and ease of handling immobilized enzymes plus advances in electronics led to the development of so-called bioelectrochemical sensors, using immobilized biocatalysts. The extension of such methods to immobilized whole cells is a natural development, using both viable and non-viable cell systems. Obviously, the latter will approach the type of applications found for individually purified and immobilized enzymes, cutting the expense of biocatalyst preparation at the risk of loss of specificity. Certain important advantages may, however, result from the use of cells. Firstly, there is the possibility of *in situ* coenzyme regeneration. Secondly, there is the additional bioselectivity and reactivity of the membrane systems of the cell, which can be exploited to advantage. Thirdly, with viable cell systems there is the possibility of regenerating the biocatalyst *in situ*, without the need to reconstruct the probe. The use of the generic term 'biosensors' has unfortunately crept into common usage to cover both immobilized enzyme and cell electrodes. The term is also used in another sense by animal physiologists concerned with directly sensing the reactions of living organisms and then includes many sensors which make no use of immobilized enzymes or cells. A more precise terminology is therefore preferable for ease of information retrieval, but the generic term 'biosensors' is likely to remain in use.

The use of cells or organelles to detoxify biological fluids is as old as multicellular life itself. Modern technology is now beginning to allow the use of immobilized cells in those medical situations and conditions where the natural system is unable, for one reason or another, to cope with toxic

217

substances. The development of sewage and wastewater treatment plants may be seen as a grosser example of detoxification, acting on a rather different type of fluid, but will not be further considered here. Brief mention does, however, need to be made of systems involving immobilized cells for the specific detoxification of pollutants.

Knowledge concerning the nature and reactivity of the cell surface, particularly of animal cells, has grown at a truly remarkable rate. The very specific affinity of the cell surface for particular reactive groups has led to a whole series of developments where the term cell immobilization correctly defines the process taking place. Only one or two aspects of this enormous subject can be touched upon in the present chapter.

Immobilized cell electrodes

Several reviews of the construction and uses of immobilized cell electrodes have appeared. Mattiasson (1979), and Suzuki & Karube (1981) considered the application of such electrodes to the analysis of a variety of substances. This was extended by Mattiasson (1983) to include a discussion of the experimental methods used. Wingard, Katchalski-Katzir & Goldstein (1981) considered analytical applications of immobilized enzymes and cells.

A critical feature of any system is the nature of the transducer used to convert the expression of the target biocatalytic activity into an electrical signal which can then be appropriately processed. In the simplest type of system an already existing electrode, such as one measuring oxygen, conductivity or specific ions (Figure 8.1), is merely covered with free cells which are then held in place with a selectively permeable membrane. Such a system is only one step removed from those where the cells, either free or immobilized, are placed separately into a reaction vessel which also contains the electrode. Matsunaga, Karube & Suzuki (1978), for example, used immobilized *Lactobacillus arabinosus* cells to estimate nicotinic acid, inserting a combined glass electrode to measure the potential change caused by lactic acid production, after one hour's incubation. The evolution from this type of spatially separate immobilized cells and sensor system, to one in which the immobilized cells are attached to the sensor is described, for BOD (biological oxygen demand) determinations, by Suzuki & Karube (1981). In this case an oxygen electrode was used as the transducer, although they also discuss a system based on a fuel cell containing *Clostridium butyricum*. Several researchers have taken tissue slices, which consist of naturally immobilized cells, and placed these directly over electrodes, held in place by a suitable membrane. Schubert, Wollenberger & Scheller (1983), for example, cut thin slices from sugar beet (*Beta vulgaris altissima*) and placed these directly on the polyethylene membrane of an oxygen

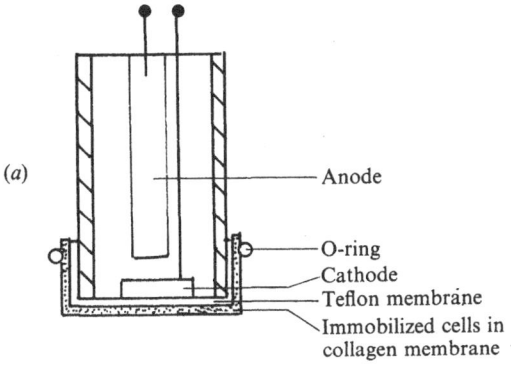

(a) Anode

O-ring
Cathode
Teflon membrane
Immobilized cells in
collagen membrane

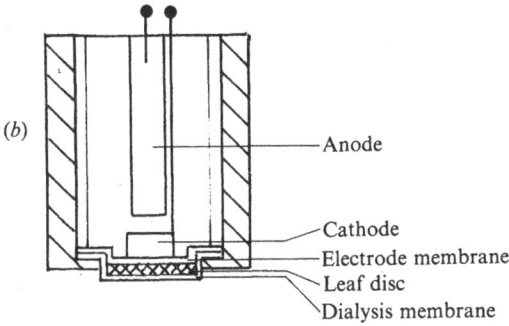

(b) Anode

Cathode
Electrode membrane
Leaf disc
Dialysis membrane

Figure 8.1. Diagrammatic representations of two typical types of immobilized cell electrodes. (*a*) BOD sensor, based on an oxygen electrode (after Suzuki and Karube, 1978); (*b*) L-cysteine leaf disc–ammonia electrode based sensor (after Smit & Rechnitz, 1984).

electrode, holding the slices in place with a piece of dialysis membrane and an O-ring. This tissue is relatively homogeneous and contains a high tyrosinase activity. A linear response to added tyrosine was obtained up to approximately 500 μmol/l, with a calibration graph allowing the range to be extended to 900 μmol/l. The tissue slices were 100 μm thick and the response time of the order of five to ten minutes. The response of the electrode system was remarkably stable at toom temperature. After an initial increase over the first two days, it remained stable for a further five days before starting to increase again. It was suggested that this might be due to permeability changes in the plant tissue. The system was also much more selective for tyrosine, as against various other tested phenolics, than had previously been reported for tyrosinase from other plant and microbial sources. This may have been due to the plasma membrane acting as a selectively permeable barrier to other phenolics. A similar principle was used by Sidwell & Rechnitz (1985) for their 'bananatrode' to be used for dopamine sensing. They used banana fruit slices held against the gas-permeable membrane of an oxygen electrode by dialysis membrane.

The simplicity of such systems makes them of great potential interest and their use is not restricted to plant tissues. Meyerhoff (1980) used slices of pig kidney, attached to a gaseous ammonia sensor, to estimate glutamine. It is doubtful, however, whether the comparable animal tissue systems will prove as satisfactory as those based on plant tissues, because of the difficulty of obtaining reproducible starting material for the preparation of the slices and the much greater instability and susceptibility to gross overgrowth by contaminating micro-organisms when using such non-sterile systems.

One advantage of the use of immobilized cells in contact with the transducer is the high cell density which can be obtained, allowing quicker detection of cellular response than can be achieved with free cells in suspension. There may, however, be an optimum cell density above which sensitivity is reduced because of cell to cell interactions. Such a situation was found by Matsunaga *et al.* (1981) when using *Leuconostoc mesenteroides* to estimate phenylalanine.

The examples already mentioned cover the extremes of the possible applications, from the broadest possible assay of any metabolizable organic compound in the BOD sensor to the specific determination of a single substance. In general, cells show a broad spectrum of enzymic activity, and the special selection of particular cells plus the development of probe configurations which limit interference from unwanted compounds must be undertaken. The use of membranes permeable to gas but not liquids is an obvious approach. Teflon (Trade Mark of Du Pont), a typical gas-permeable membrane was used to separate immobilized *Trichosporon brassicae* from the external solution in an oxygen electrode-based system to assay acetic acid (Hikuma *et al.*, 1979). The pH value of the external solution must be kept below the pK value of acetic acid because only the molecular form of acetic acid and not the acetate ion can pass through the Teflon membrane. The membrane effectively excludes all non-volatile components of the medium which, under direct contact, could be used for respiration by the *T. brassicae*. Certain other substances do, however, pass the membrane and it has even been used, under appropriate operating conditions (neutral pH values) to measure ethanol. With acetic acid the range of concentrations which give a linear response was 5–72 mg/l and the correlation coefficient between the sensor determinations and a gas chromatographic determination was 1.04. The system was stable over three weeks of use, involving 1500 separate assays. For ethanol determinations the correlation coefficient with gas chromatography was 0.98 with good stability over a similar time period and over 2000 assays. Where such simple physical methods are not possible it may be necessary to resort to the selection of micro-organisms with stringent nutritional requirements. Such information is, of course, readily available for many species because it forms the basis of much of the species discrimination of bacteria. Another ready source of such special characteristics is to be found in the ranks of the many thousands of mutant micro-organisms which are available and, as yet, essentially untapped for use in this type of sensor.

So far the sensors mentioned have made use of the promotion of metabolic activity produced by the substrate being assayed. In some situations, however, it is a reduction in metabolic activity which is measured. The quantitative assay of antibiotic concentrations is of paramount importance to many pharmaceutical companies. The possibility of detecting such substances with direct-reading sensors has therefore received some attention. Suzuki & Karube (1979a) immobilized cells of *Citrobacter freundii* in a collagen membrane. This bacterium can hydrolyse cephalosporin, with the liberation of hydrogen ions which were detected using a combined glass electrode. In this case the pH electrode was physically separate from the immobilized cells. The cephalosporinase activity of the cells was active for at least a month and the results compared favourably with those obtained using conventional HPLC (high performance liquid chromatography) of fermentation broths. In the same paper Suzuki & Karube also described an immobilized cell reactor for glucose, based on *Pseudomonas fluorescens* in collagen membranes. In this case the immobilized cells were directly placed onto a gas-permeable Teflon membrane covering a platinum/lead electrode system. The system was highly selective for glucose, giving less than 10% relative response to galactose and less than 5% response to fructose, mannose or sucrose. Again, a stability of at least a month was claimed for the immobilized bacteria.

The antifungal substance nystatin has been assayed using yeast cells immobilized by adsorption onto filter paper and an oxygen electrode. This made specific use of the observation that the current produced was reduced as the cell number in the electrode membrane was increased. By finding the range of cell numbers over which the relationship was linear, and starting with a cell number giving a minimal reading, it was possible to relate cell death to increase in current product. Although this may appear a cumbersome approach, the preparation of the immobilized cell membranes and the changing of a new membrane for an old one were both simple and rapid. Clearly there are many possible approaches to the detection of antibiotics.

The general detection of toxic substances can also be accomplished by the use of immobilized cells. Mattiasson (1979) described these as 'poison guards' and emphasized their value in situations where the total toxic effect is more important than the specific identification and quantification of individual toxic substances. One system which was tried for this purpose consisted of bacteria immobilized onto rotating plates and followed by an oxygen electrode (Solyom, 1976). This could detect sudden harmful changes in wastewaters by the rise in dissolved oxygen and thereby allow the main filter beds to be protected from damage by diversion of the toxic waste stream.

An interesting detection system for mutagens has been described by Karube *et al.* (1981). Two identical oxygen electrodes were covered with porous acetyl cellulose membranes containing *Bacillus subtilis*. One probe was made up with the normal wild-type strain which is competent in the repair of mutagen-damaged DNA (REC^+ in genetical notation). The other probe contains a strain deficient in this repair system (REC^-). Both electrodes are inserted into a single test vessel containing the suspected mutagen. The REC^- strain shows greater cell death than the REC^+ and this is quantified by comparison of the electrodes. The method was more sensitive than the Ames test and gave a response in as little as one hour, with no difference being found between the two strains when non-mutagenic inhibitors of metabolism were tested.

Apart from the detection of the chemical products of metabolism it is also possible to detect the heat produced as a result of the inefficiency of metabolic transformations. A sensitive thermistor can be inserted into an immobilized cell reactor, suitably thermally insulated, and the influence of added substrates can be easily quantified. With suitable modifications the system can operate on a continuous flow system as described by Mattiasson, Larsson & Mosbach (1977). As with other non-specific transducers the thermistor suffers from complete lack of specificity and the biocatalyst must be carefully chosen and used under very controlled conditions. The measurement of small changes in heat content also requires rather sophisticated equipment which is not generally available in most laboratories. The same difficulty of access applies also to probes based on membrane inlet mass spectrometry. This very powerful technique can be used to identify and quantify volatile substances directly and is being developed in its own right as a means of monitoring the progress of fermentations. Coupled with immobilized cells the range of applications could be further extended.

Maisterrena, Blum & Coulet (1986) carried out a mathematical analysis of product transfer through a permeable enzyme membrane. They concluded that a product hyperconcentration effect could be obtained if sequentially acting enzymes were grafted onto opposite sides of a membrane, with the final stage on the side next to the smaller (inner) compartment. This theoretical analysis is of direct interest in the development of multi-enzyme biosensors.

Detoxification units

As legislation to protect the environment and the human and animal food chains from pollution expands, attention increasingly turns to the removal of toxic substances from effluent streams. Some of these merely use biological material as a cheap alternative to selective ion-exchange resins, making use of the charged nature of the cell wall to promote the removal of specific charged ions. Concentration factors for uptake into the cellular material may be several thousand-fold for metallic cations and the ions can, if required, be recovered by elution with acid, at the same time regenerating the active form of the adsorbant. For a discussion of these methods the reader is referred to the extensive literature on wastewater treatment and the prevention of heavy metal pollution.

As an example of a system using conventional calcium alginate immobilization the research of Nilsson *et al.* (1980) can be considered. The presence of residual nitrate and/or nitrite in drinking water is a potential health hazard. These anions arise largely as a result of excessive use of soluble forms of nitrogenous fertilizers, which are washed out of the soil but are not removed by the conventional treatments given to potable waters. Complete removal can, in theory, be obtained by the use of denitrifying bacteria to convert the inorganic forms of nitrogen into gaseous nitrogen, which is then lost to the atmosphere. Viable *Pseudomonas denitrificans* were immobilized in bead or fibre form using calcium alginate. Denitrification is an anaerobic process but the bacteria must be supplied with an appropriate source of organic carbon. The nature of the carbon source influences the rate of denitrification and sodium aspartate was chosen, on the basis of trial experiments with free, non-growing *P. denitrificans*. As expected, the size of the immobilization gel particles influenced the rate of denitrification, probably due to diffusion limitation. Maintaining gel integrity in a situation where gas bubbles are continually being formed necessitated operation in the presence of calcium ions while particle size had to be kept sufficiently large to prevent adverse effects on the flow rate of the packed-bed column studied. This was overcome by using fibres rather than beads. As expected, apparent K_m values for the nitrate reductase activity of immobilized cells (0.97 mM) were considerably increased over those for the free cell (0.021 mM). An approximate half-life of 30 days was estimated but regeneration of the active biomass was possible. Improvements suggested were to enhance mechanical stability of the gel and reduce the cost of the added carbon source.

Chevalier & De La Noüe (1986) used carrageenan-immobilized *Scenedesmus quadricauda* to remove ammonium and orthophosphate from solution and proposed the possible use of hyperconcentrated microalgae for tertiary wastewater treatment. There are many other interesting possibilities if economically viable systems can be developed. The use of microbial cells for effluent clean-up is not restricted to organic and heavy metal pollutants. Macaskie & Dean (1985b) for example, used polyacrylamide-immobilized *Citrobacter* to remove strontium from solution and suggested its possible application in the removal of radioactive strontium from waste streams at reprocessing plants. *Citrobacter* was also used by Macaskie & Dean (1985a) for uranium removal.

As an example of a direct medical detoxification system the work of Kastl *et al.* (1979) can be considered. They investigated the use of rat liver microsomes entrapped in a hollow-fibre system. Before extraction the hydroxylation enzyme activity of the microsomes was induced by phenobarbitol treatment. The advantage of using microsomes, rather than purified immobilized enzymes, is the greater range of substrates detoxified by the microsomes, which function as multi-enzyme complexes. The two model drugs were *p*-nitroanisole and hexobarbitol. The microsome preparation could be kept for over 14 months at 23 °C, freeze-dried and stored in evacuated containers. The system required a constant supply of active coenzyme which involved both addition of exogenous coenzyme and a suitable substrate for its regeneration. When human plasma was added the rate of hexobarbitol metabolism was reduced by approximately 50%, due, it was suggested, to binding of the drug to plasma proteins and blocking of the membrane pores. The steady-state rates of hexobarbitol metabolism were similar for free microsomes and those entrapped behind membranes having nominal molecular weight cut-offs of 30 000 and 50 000. Extrapolation of the results suggested that the microsome system was much slower than conventional charcoal haemoperfusion but approached that of a haemodialysis. Despite this low efficiency it was felt that the metabolic capability of the immobilized organelles, coupled with the possibility of developing a more efficient system capable of perhaps providing a liver support system, made further research worthwhile. This one example serves to illustrate the magnitude of the difference between the rather crude systems suitable for detoxifying drinking water and medical detoxification units, with their enormously more stringent operational characteristics.

Cell affinity immobilization

Affinity immobilization makes use of the unique surface characteristics of cells to allow the selective binding of a subpopulation of cells from a mixed population. It therefore differs fundamentally in concept from other types of immobilization, where the bonding is non-selective, allowing the same method to be used for many, widely different, cell types. For obvious reasons, affinity immobilization is sometimes described as biospecific immobilization. In general, the binding involves the plasma membrane and the method is therefore particularly applicable to animal cells, but it has also been used fairly extensively for protoplasts. There are two major ways of using affinity immobilization. One involves carrier beads and the other has reactive groups firmly attached to the walls of an appropriate glass or plastic reaction chamber. The beaded carriers may be added to a mixed cell population, allowed to react and then decanted and washed free of unbound cells. Alternatively, the beads may be packed into a column and the system operated as a cell affinity chromatography column. The affinity chromatography of macromolecules is, of course, a well-established branch of biochemical separation methodology. A discussion of recommended terminology is found in Pye & Weetall (1975). Where a reaction vessel has been prepared for cell affinity immobilization it may take any desired form. Syringes can be internally coated so that certain cell types are retained inside the syringe when it is rinsed. Tubes, bottles and micro-titre wells may be coated and used in a similar way.

The essential feature of affinity immobilization is that the cells self-attach to a ligand which is covalently joined to the support material, whatever that might be (Figure 8.2). If the individual cells are bound at only one or two points to the ligands, it is likely that they can be easily removed by elution from the support with an appropriate substance that competes for the ligand binding site. If multiple point binding of the individual cells to many ligand binding sites occurs it may be impossible to remove the cells by this competitive elution technique. For some purposes, such as cell affinity chromatography, it may be essential to recover the cells again in a viable state and so relatively weak immobilization is sought. In an industrial process weak binding would allow the removal of non-functional cells and the recovery of the expensive activated support material for recharging with culture cells. Ligands could be bound to almost any support material. One interesting example of this is the use of activated magnetite particles specifically to immobilize cancerous cells from body fluids. These particles can then be held in a magnetic field and the normal cells in the body fluid returned to the patient, while the cancerous cells are removed and destroyed. This, in fact, represents a particular form of 'detoxification' unit, removing whole cells rather than molecular toxic substances. Probably the most commonly used support materials for the affinity immobilization of animal cells, however, are carbohydrate-based. A convenient size of bead is between 250 and 350 μm, this giving sufficiently large gaps between the beads to allow easy washing out of cells during affinity chromatography.

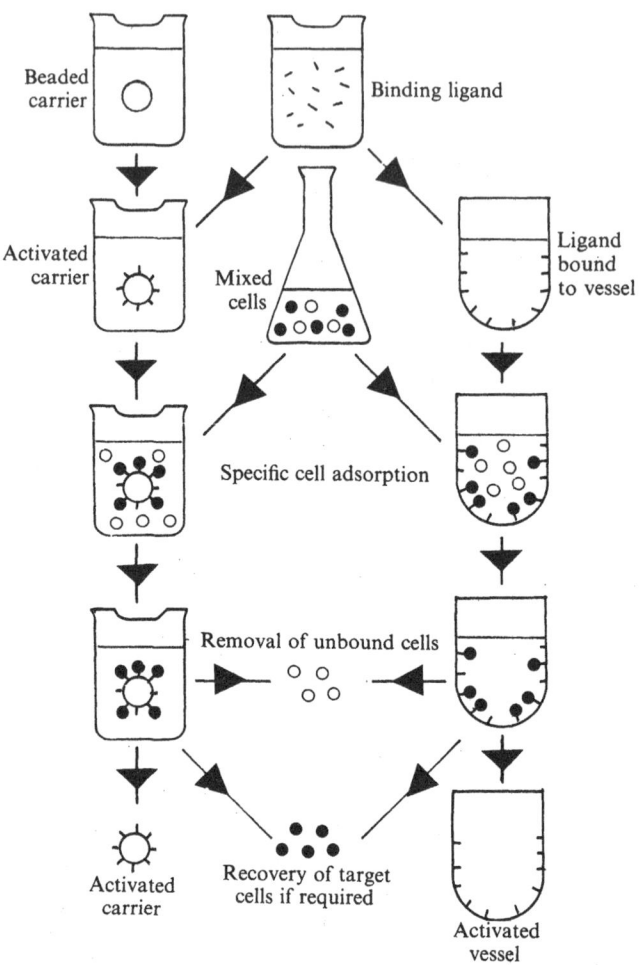

Figure 8.2. Diagrammatic representation of the principle of cell affinity immobilization.

Three major types of interaction are possible. Firstly, the highly specific nature of the antibody–antigen reaction can be utilized. The purified antibody preparation is covalently linked to the support, for example using a CNBr-activated support. When a mixed cell population is added to this, those cells carrying the appropriate antigen are bound, leaving the free cell population specifically depleted of the bound cell type. Marshak-Rothstein *et al.* (1979) used monoclonal antibodies specific for Thy-1 antigens of mouse T-cell lymphocytes, bound to Sepharose 6MB (Trade Mark of Pharmacia Fine Chemicals), to fractionate mouse lymph node cells. This gave a residual free cell population containing not more than 2% of the affinity target T-cells. The T-cells themselves were recovered in this research by 10 seconds of mechanical vortexing and they showed 90% viability. Because of the difficulty of recovering the bound cells in good condition an alternative strategy has been developed. This uses an affinity support to which the so-called Protein A has been covalently bound. Protein A, from *Staphylococcus aureus* binds IgG protein via the Fc region, which is distinct and separate from the antigen-binding site of the antibodies. If the target cells are first reacted with free specific antibodies, these cell-bound antibodies can then be, in their turn, bound via the Fc region to the affinity support. When required, the bound cells can be easily and gently removed from the support by competitive elution using excess free IgG.

The second type of interaction involves the use of covalently bound lectins, which bind with certain carbohydrate residues. Generally, the carbohydrates are part of the glycoproteins found in the plasma membrane, but they could also form part of a cell wall. Generally the cells have several different sugar residues exposed and it may be difficult to find a system which discriminates between sub-populations, although binding of a whole population may be relatively easy. There is a wide range of possible lectins known, with differing sugar specificities, but only a few have been used extensively for affinity immobilization. Many are derived from plant sources. Bound cells can be released easily from the supports by competitive elution with an appropriate soluble sugar. Supports are available with CNBr-activation for the users to attach their own lectins or with specific lectins, such as that from wheat germ, already attached. A problem which may occur with some lectins is that they may react directly with carbohydrate-based support materials. This problem may be overcome, in a rather elaborate way, by generating antibodies to the lectin and immobilizing the antibodies instead. The lectin may then be bound separately to the receptor cells and the cell–lectin complex is then capable of attaching to the antibody-coated support. The appropriate sugar can then be used in a cell-elution medium. This method was used by Irle, Piguet & Vassalli (1978) with mouse thymocytes and peanut lectin. A number of examples of cell affinity immobilization are described in the trade booklets of affinity support material manufacturers (for example, Pharmacia Fine Chemicals, 1980).

The third type of affinity carrier makes use of specific receptor sites on the cell surface. These might be sites for hormones, host-specific toxins or some other effector molecule. Provided that the appropriate active substance is available in sufficient quantity, it can be attached to the support, for example, using a CNBr-activated support. After binding. the cells may be released by a specific inhibitor of the receptor–ligand interaction or often by using excess free ligand.

The existence of receptor sites on the plasma membranes of plant cells is also well-known. Larkin (1981) reviewed the general subject of protoplast agglutination and immobilization. Free protoplasts of many plant species are agglutinated by free multivalent lectins derived from other (unrelated) plant species. Although this might be seen as a form of assisted flocculation, it is of fundamental rather than currently practical significance because any form of flocculation will tend to trap passively other (non-reacting) free cells, which will contaminate the preparation. Lectins have also been used, covalently bound to protein supports, to bind plant protoplasts. Host-specific toxins from plant pathogens can also be used as binding ligands. This could lead to a relatively simple method of selecting resistant mutants from populations of plant protoplasts derived from susceptible host cell lines. The extreme specificity of cell affinity immobilization suggested many possible applications.

References

Aaronson, S. (1981). *Chemical Communication at the Microbial Level*, vol. 1. Boca Raton: CRC Press.

Adamson, S. R., Fitzpatrick, S. L., Behie, L. A., Gaucher, G. M. & Lesser, B. H. (1983). *In vitro* production of high titre monoclonal antibody by hybridoma cells in dialysis culture. *Biotechnology Letters*, **5**, 573–8.

Adlercreutz, P. (1986). Oxygen supply to immobilized cells: Theoretical calculations and experimental data for the oxidation of glycerol by immobilized *Gluconobacter oxydans* cells with oxygen or *p*-benzoquinone as electron acceptor. *Biotechnology and Bioengineering*, **28**, 223–32.

Ahern, T. J., Kator, S. & Sada, E. (1983). Prostaglandin synthesis from arachidonic acid by immobilized ram seminal microsomes. *Biotechnology and Bioengineering*, **25**, 881–5.

Aiba, S., Humphrey, A. E. & Millis, N. F. (1973). *Biochemical Engineering*, 2nd edn. New York: Academic Press.

Albertsson, P. A. (1971). *Partition of Cell Particles and Macromolecules*, 2nd edn. New York: Wiley Interscience.

Altshuler, G. L., Dziewulski, D. M., Sowek, J. A. & Belfort, G. (1986). Continuous hybridoma growth and monoclonal antibody production in hollow fiber reactors-separators. *Biotechnology and Bioengineering*, **28**, 646–57.

Amicon Corporation (1985). Unique hollow fiber cell culture system. *Technical Data Publication 482*. Danvers: Amicon Corporation.

Anderson, J. G., Blain, J. A., Divers, M. & Todd, J. R. (1980). Use of the disc fermenter to examine production of citric acid by *Aspergillus niger*. *Biotechnology Letters*, **2**, 99–104.

Anderson, J. G., Blain, J. A., Marchetti, P. & Todd, J. R. (1981). Processing of model dilute carbohydrate wastes using *Aspergillus niger* in disc fermenters. *Biotechnology Letters*, **3**, 451–4.

Andrews, G. F. & Przezdziecki, J. (1986). Design of fluidized-bed fermentors. *Biotechnology and Bioengineering*, **28**, 802–10.

Antrim, R. L., Colilla, W. & Schnyder, B. J. (1979). Glucose isomerase production of high-fructose syrups. In *Applied Biochemistry and Bioengineering*, vol. 2, *Enzyme Technology*, ed. L. B. Wingard, E. Katchalski-Katzir & L. Goldstein. New York: Academic Press.

Antonioni, E., Carrera, G. & Cremonesi, P. (1981). Enzyme catalyzed reactions in water-organic solvent two-phase systems. *Enzyme and Microbial Technology*, **3**, 291–6.

Arcuri, E. J., Slaff, G. & Greasham, R. (1986). Continuous production of thienamycin in immobilized cell systems. *Biotechnology and Bioengineering*, **28**, 842–9.

Arcuri, E. J., Worden, R. M. & Shumate, S. E., II. (1980). Ethanol production by immobilized cells of *Zymomonas mobilis*. *Biotechnology Letters*, **2**, 499–504.

229

Arkles, B. & Brinigar, W. S. (1975). Respiratory properties of rat liver mitochondria immobilized on an alkylsilylated glass surface. *Journal of Biological Chemistry*, **250**, 8856–62.

Artavanis, G. & Todd, J. R. (1980). Sulphite oxidation in the disc fermenter. *Biotechnology Letters*, **2**, 23–8.

Atkinson, B. (1974). *Biochemical Reactors*. London: Pion Limited.

Atkinson, B., Black, G. M. & Pinches, A. (1980). Process intensification using cell support systems. *Process Biochemistry*, **15**, 24–32.

Banerjee, M., Chakrabarty, A. & Majumdar, S. K. (1982). Immobilization of yeast cells containing β-galactosidase. *Biotechnology and Bioengineering*, **24**, 1839–50.

Bang, W.-G., Behrendt, U., Lang, S. & Wagner, F. (1983). Continuous production of L-tryptophan from indole and L-serine by immobilized *Escherichia coli* cells. *Biotechnology and Bioengineering*, **25**, 1013–25.

Baratti, J., Varma, R. & Bu'Lock, J. D. (1986). High productivity ethanol fermentation on a mineral medium using a flocculent strain of *Zymomonas mobilis*. *Biotechnology Letters*, **8**, 175–80.

Bell, G., Blain, J. A., Patterson, J. D. E., Shaw, C. E. L. & Todd, R. (1978). Ester and glyceride synthesis by *Rhizopus arrhizus* mycelia. *Federation of European Microbiological Societies Microbiology Letters*, **3**, 223–5.

Bennetto, H. P., Delaney, G. M., Mason, J. R., Roller, S. D., Stirling, J. L. & Thurston, C. F. (1986). The sucrose fuel cell: efficient biomass conversion using a microbial catalyst. *Biotechnology Letters*, **7**, 699–704.

Bennetto, H. P., Stirling, J. L., Tanaka, K. & Vega, C. A. (1983). Anodic reactions in microbial fuel cells. *Biotechnology and Bioengineering*, **25**, 559–68.

Birnbaum, S., Pendleton, R., Larsson, P. & Mosbach, K. (1981). Covalent stabilization of alginate gel for the entrapment of living whole cells. *Biotechnology Letters*, **3**, 393–400.

Blain, J. A., Anderson, J. G., Todd, J. R. & Divers, M. (1979). Cultivation of filamentous fungi in the disc fermenter. *Biotechnology Letters*, **1**, 269–74.

Bland, R. R., Chen, H. C., Jewell, W. J., Bellamy, W. D. & Zall, R. R. (1982). Continuous high rate production of ethanol by *Zymomonas mobilis* in an attached film expanded bed fermenter. *Biotechnology Letters*, **4**, 323–8.

Bornman, C. H., Olesen, P. & Zachrisson, A. (1983). Microcarrier-anchored plant protoplasts. In *Proceedings of the 6th International Protoplast Symposium*, ed. I. Potrykus *et al.*, pp. 270–1. Basel: Birkhauser.

Bowen, I. D. & Lockshin, R. A. (1981). *Cell Death in Biology and Pathology*. London: Chapman & Hall.

Breslau, B. R. & Kilcullen, B. M. (1975). Hollow fiber enzymatic reactors: an engineering approach. In *Enzyme Engineering*, vol. 3, ed. E. K. Pye & H. H. Weetall, pp. 179–90. New York: Plenum Press.

Bringi, V. & Dale, B. E. (1985). Enhanced yeast immobilization by nutrient starvation. *Biotechnology Letters*, **7**, 905–8.

Brink, L. E. S. & Tramper, J. (1986a). Modelling the effects of mass transfer on kinetics of propene epoxidation of immobilized *Mycobacterium* cells: 1. Pseudo-one-substrate conditions and negligible product inhibition. *Enzyme and Microbial Technology*, **8**, 281–8.

Brink, L. E. S. & Tramper, J. (1986b). Modelling the effects of mass transfer on kinetics of propene epoxidation of immobilized *Mycobacterium cells: 2. Product inhibition. Enzyme and Microbial Technology*, **8**, 334–40.

Broad, D. F., Foulkes, J. & Dunnill, P. (1984). The uptake of *Aspergillus ochraceus* spores on diatomaceous particles and their use in the 11-α-hydroxylation of progesterone. *Biotechnology Letters*, **6**, 357–62.

Brodelius, P. & Mosbach, K. (1982). Immobilized plant cells. *Advances in Applied Microbiology*, **28**, 1–26.

Brodelius, P. & Nilsson, K. (1980). Entrapment of plant cells in different matrices. *Federation of European Biochemical Societies Letters*, **122**, 312–6.

Broun, G. B., Manecke, C. & Wingard, L. M. (1978). Guidelines for the characterization of immobilized enzymes: a proposal. In *Enzyme Engineering*, vol. 4, pp. 463–5. New York: Plenum Press.

Buchholz, K. (1979). *Characterization of Immobilized Biocatalysts*. New York: Verlag Chemie.

Buchholz, K. & Godelmann, B. (1979). Pressure drop and flow resistance in fixed bed reactors. In *Characterization of Immobilized Biocatalysts*, ed. K. Buchholz, pp. 127–35. New York: Verlag Chemie.

Bucke, C. (1983). Immobilized cells. *Philosophical Transactions of the Royal Society, London, Series B*, **300**, 369–89.

Bull, A. T., Holt, G. & Lilly, M. D. (1982). *Biotechnology: International Trends and Perspectives*. Paris: OECD.

Bull, M. P., Sterritt, R. M. & Lester, J. N. (1983). An evaluation of four start-up regimes for anaerobic fluidized bed reactors. *Biotechnology Letters*, **5**, 333–8.

Burstein, C., Ounissi, H., Legoy, M. D., Gellf, G. & Thomas, D. (1981). Recycling of NAD using coimmobilized alcohol dehydrogenase and *E. coli*. *Applied Biochemistry and Biotechnology*, **6**, 329–38.

Calam, C. T., Driver, N. & Bowers, R. H. (1951). Studies in the production of penicillin, respiration and growth of *Penicillium chrysogenum* in submerged culture, in relation to agitation and oxygen transfer. *Journal of Applied Chemistry*, **1**, 209–16.

Callander, I. J. & Barford, J. P. (1983). Cheese whey anaerobic digestion – effect of chemical flocculant addition. *Biotechnology Letters*, **5**, 153–8.

Chang, H. N., Joo, I. S. & Ghim, Y. A. (1984). Performance of rotating packed disk reactor with immobilized glucose oxidase. *Biotechnology Letters*, **6**, 487–92.

Cheetham, P. S. J. (1980). Developments in the immobilisation of microbial cells and their applications. In *Topics in Enzyme and Fermentation Biotechnology*, vol. 4, ed. A. Wiseman, pp. 189–238. Chichester: Ellis Horwood Ltd.

Chen, H. C. & Zall, R. R. (1982). Continuous fermentation of whey into alcohol using an attached film expanded bed reactor. *Process Biochemistry*, **17**(1), 20–5.

Cheryan, M. & Mehaia, M. A. (1983). A high-performance membrane bioreactor for continuous fermentation of lactose to ethanol. *Biotechnology Letters*, **5**, 519–24.

Chevalier, P. & De La Noüe, J. (1986). Efficiency of immobilized hyperconcentrated algae for ammonium and orthophosphate removal from waste-waters. *Biotechnology Letters*, **7**, 395–400.

Chibata, I. (1979). Immobilized microbial cells with polyacrylamide gel and carrageenan and their industrial applications. In *Immobilized Microbial Cells, ACS Symposium Series 106*, ed. K. Venkatsubramanian, pp. 187–202. Washington DC: American Chemical Society.

Chibata, I. & Tosa, T. (1976). Industrial applications of immobilized enzymes and immobilized microbial cells. In *Applied Biochemistry and Bioengineering*, vol. 1, *Immobilized Enzyme Principles*, ed. L. B. Wingard, E. Katchalski-Katzir & L. Goldstein, pp. 329–57. New York: Academic Press.

Chibata, I. & Tosa, T. (1981). Use of Immobilized Cells. *Annual Reviews of Biophysics and Bioengineering*, **10**, 197–216.

Chibata, I. & Wingard, L. B., Jr (1983). *Applied Biochemistry and Bioengineering*, vol. 4, *Immobilized Microbial Cells*. New York: Academic Press.

Chilver, M. J., Harrison, J. & Webb, T. J. B. (1978). Use of immunofluorescence and viability stains in quality control. *Journal of the American Society of Brewing Chemistry*, **36**, 13–18.

Clarke, A. R. & Forster, C. F. (1982). Biopolymer yields from activated sludge and their relation to the operation of treatment plant. *Biotechnology Letters*, **4**, 655–60.

Cocquempot, M. E., Aguirre, R., Lissolo, T., Monsan, P., Hatchikian, E. C. & Thomas, D. (1982). Co-immobilization effect on H_2 production by a chloroplast membranes–hydrogenase system. *Biotechnology Letters*, **4**, 313–18.

Cooper, P. F. (1981). The use of biological fluidized beds for the treatment of domestic and industrial wastewaters. *The Chemical Engineer*, **371/2**, 373–6.

Costerton, J. W. & Cheng, K. J. (1982). Microbe–microbe interactions at surfaces. In *Experimental Microbial Ecology*, ed. R. G. Burns & J. H. Slater, pp. 275–90. Oxford: Blackwell Scientific Publishers.

Dainty, A. L., Goulding, K. H., Robinson, P. K., Simpkins, I. & Trevan, M. D. (1986). Stability of alginate-immobilized algal cells. *Biotechnology and Bioengineering*, **28**, 210–16.

Damiano, D. & Wang, S. S. (1986). Novel use of perfluorocarbon for supplying oxygen to aerobic submerged cultures. *Biotechnology Letters*, **7**, 81–6.

Daugulis, A. J., Brown, N. M., Cluett, W. R. & Dunlop, D. B. (1981). Production of ethanol by adsorbed yeast cells. *Biotechnology Letters*, **3**, 651–6.

De Cabrera, S., De Arriola, M. C., Morales, E., De Micheo, F. & Rolz, C. (1981). Ex-Ferm ethanol production in packed bed fermentors: A three cycle operation employing chipped sugarcane. *Biotechnology Letters*, **3**, 497–502.

Denac, M., Uzman, S., Tanaka, H. & Dunn, I. J. (1983). Modeling of experiments on biofilm penetration effects in a fluidized bed nitrification reactor. *Biotechnology and Bioengineering*, **25**, 1841–61.

Deo, Y. M. & Gaucher, G. M. (1983). Semi-continuous production of the antibiotic patulin by immobilized cells of *Penicillium urticae*. *Biotechnology Letters*, **5**, 125–30.

De Rosa, M., Gambacorta, A., Nicolaus, B., Buonocore, V. & Poerio, E. (1980). Immobilized bacterial cells containing a thermostable β-galactosidase. *Biotechnology Letters*, **2**, 29–34.

De Rosa, M., Gambacorta, A., Lama, L. & Nicolaus, B. (1981). Immobilization of thermophilic microbial cells in crude egg white. *Biotechnology Letters*, **3**, 183–6.

Deshpande, V., SivaRaman, H. S. & Rao, M. (1983). Simultaneous saccharification and fermentation of cellulose to ethanol using *Penicillium funiculosum* cellulase and free or immobilized *Saccharomyces uvarum* cells. *Biotechnology and Bioengineering*, **25**, 1679–84.

Dias, S. M. M., Novais, J. M. & Cabral, J. S. (1982). Immobilization of yeasts on titanium activated inorganic supports. *Biotechnology Letters*, **4**, 203–8.

Dinelli, D. (1972). Entrapment in solid fibres. *Process Biochemistry*, **7(8)**, 9–12.

DiSpirito, A. A., Dugan, P. R. & Tuovinen, O. H. (1983). Sorption of *Thiobacillus ferrooxidans* to particulate material. *Biotechnology and Bioengineering*, **25**, 1163–8.

Dix, P. J., Kane, E. J., Keane, G. J. & O'Sullivan, M. T. (1983). Immobilisation of *Nicotiana* cells and protoplasts in alginate beads. In *6th International Protoplast Symposium*, ed. I. Potrykus *et al.*, pp. 226–7. Basel: Birkhauser.

Doran, P. M. & Bailey, J. E. (1986). Effects of immobilization on growth fermentation properties and macromolecular composition of *Saccharomyces cerevisiae* attached to gelatin. *Biotechnology and Bioengineering*, **28**, 73–87.

D'Souza, S. F., Melo, J. S., Deshpande, A. & Nadkarni, G. B. (1986). Immobilization of yeast cells by adhesion to glass surface using polyethylenimine. *Biotechnology Letters*, **8**, 643–8.

Duff, S. J. B. & Kennedy, K. J. (1982). Effect of hydraulic and organic overloading on thermophilic downflow stationary fixed film (DSFF) reactor. *Biotechnology Letters*, **4**, 815–20.

Duff, S. J. B. & Kennedy, K. J. (1983). Effect of effluent recirculation on start-up and steady state operation of the downflow stationary fixed film (DSFF) reactor. *Biotechnology Letters*, **5**, 317–20.

Egerer, P. & Simon, H. (1982). Hydrogenation with gel entrapped *Clostridium* sp. La 1 and observation on its stability. *Biotechnology Letters*, **4**, 501–6.

Egerer, P., Simon, H., Tanaka, A. & Fukui, S. (1982). Immobilization and stability of the NAD dependent hydrogenase from *Alcaligenes eutrophus* and of the whole cells. *Biotechnology Letters*, **4**, 489–94.

Einsele, A. (1978). Scaling up bioreactors. *Process Biochemistry*, **13**, 13–14.

Ellwood, D. C., Keevil, C. W., Marsh, P. D., Brown, C. M. & Wardell, J. N. (1982). Surface-associated growth. *Philosophical Transactions of the Royal Society, London, Series B*, **297**, 517–32.

Enfors, S. O. & Mattiasson, B. (1983). Oxygenation of processes involving immobilized cells. In *Immobilized Cells and Organelles*, vol. II, ed. B. Mattiasson, pp. 41–60. Boca Raton: CRC Press.

Engasser, J. M. & Horvath, C. (1976). Diffusion and kinetics with immobilized enzymes. In *Applied Biochemistry and Bioengineering*, vol. 1, ed. L. B. Wingard, Jr., E. Katchalski-Katzir & L. Goldstein, pp. 127–220. New York: Academic Press.

Fein, J. E., Lawford, H. G., Lawford, G. R., Zawadzki, B. C. & Charley, R. C. (1983). High productivity continuous ethanol fermentation with a flocculating mutant strain of *Zymomonas mobilis*. *Biotechnology Letters*, **5**, 19–24.

Felix, H. R. (1980). Thesis presented for degree of Doctor of Philosophy, Department of Microbiology, University of Basel. Basel: University of Basel.

Felix, H., Brodelius, P. & Mosbach, K. (1981). Enzyme activities of the primary and secondary metabolism of simultaneously permeabilized and immobilized plant cells. *Analytical Biochemistry*, **161**, 462–70.

Felix, H. R. & Mosbach, K. (1982). Enhanced stability of enzymes in permeabilized and immobilized cells. *Biotechnology Letters*, **4**, 181–6.

Flaschel, E. & Wandrey, C. (1979). Membrane Reactors. In *Characterization of Immobilized Biocatalysts*, ed. K. Buchholz, pp. 337–66. New York: Verlag Chemie.

Fletcher, M. (1976). The effects of proteins on bacterial attachment to polystyrene. *Journal of General Microbiology*, **94**, 400–4.

Forster, C. F., Rockey, J. S., Wase, D. A. J. & Godwin, S. J. (1982). Mixing characteristics of fixed film anaerobic reactors. *Biotechnology Letters*, **4**, 799–804.

Freeman, A. & Aharonowitz, Y. (1981). Immobilization of microbial cells in cross-linked, prepolymerized, linear polyacrylamide gels: antibiotic production by immobilized *Streptomyces clavuligerus* cells. *Biotechnology and Bioengineering*, **23**, 27–47.

Frein, E. M., Montenecourt, B. S. & Eveleigh, D. E. (1982). Cellulase production by *Trichoderma reesei* immobilized on K-carrageenan. *Biotechnology Letters*, **4**, 287–92.

Fujimura, T. & Kaetsu, I. (1981). Immobilization of yeast cells by radiation-induced polymerization. *Zeitschrift für Naturforschung*, **37**, 102–6.

Fukui, S. & Tanaka, A. (1982). Immobilized microbial cells. *Annual Reviews of Microbiology*, **36**, 145–72.

Fukui, S., Tanaka, A. & Gellf, G. (1978). Immobilization of enzymes, microbial cells and organelles by inclusion with photo-crosslinkable resins. In *Enzyme Engineering*, vol. 4, ed. G. B. Broun, G. Maneke & L. M. Wingard, pp. 299–306. New York: Plenum Press.

Fuller, K. W. & Bartlett, D. J. (1985). The chemosynthetic potential of plants and its realisation by immobilized cells. *Annual Proceedings of the Phytochemical Society of Europe*, **26**, 229–47.

Gaden, E. L. (1955). Fermentation kinetics and productivity. *Chemistry and Industry*, **7** (12 Feb), 154–9.

Gahan, P. B. (1981). Cell senescence and death in plants. In *Cell Death in Biology and Pathology*, ed. I. D. Bowen & R. A. Lockshin, pp. 145–69. London: Chapman & Hall.

Gbewonyo, K. & Wang, D. I. I. (1983). Confining mycelial growth to porous microbeads: A novel technique to alter the morphology of non-Newtonian mycelial cultures. *Biotechnology and Bioengineering*, **25**, 967–83.

Gekas, V. C. (1986). Artificial membranes as carriers for the immobilization of biocatalysts. *Enzyme and Microbial Technology*, **8**, 450–60.

Genung, R. K., Hancher, C. W., Rivera, A. L. & Harris, M. T. (1982). Energy conservation and methane production in municipal wastewater treatment using fixed-film, anaerobic bioreactors. In *Biotechnology and Bioengineering, Symposium no. 12*, ed. E. L. Gaden, Jr, pp. 365–380. New York: John Wiley.

Gerson, D. F. & Zajic, J. E. (1979). The biophysics of cellular adhesion. In *Immobilized Microbial Cells, ACS Symposium Series 106*, ed. K. Venkatsubramanian, pp. 29–57. Washington, DC: American Chemical Society.

Gestreluis, S. (1983). Immobilized nonviable cells for use of a single or a few enzyme steps. In *Immobilized Cells and Organelles*, vol. II, ed. B. Mattiasson, pp. 1–32. Boca Raton: CRC Press.

Ghommidh, C., Navarro, J. M. & Durand, G. (1981). Acetic acid production by immobilized *Acetobacter* cells. *Biotechnology Letters*, **3**, 93–8.

Ghommidh, C., Navarro, M. M. & Messing, R. A. (1982). A study of acetic acid production by immobilized *Acetobacter* cells: Product inhibition effects. *Biotechnology and Bioengineering*, **24**, 1991–9.

Gisby, P. E. & Hall, D. O. (1980). Biocatalytic H_2 production using alginate-immobilized chloroplasts, enzymes and synthetic catalysts. *Nature*, **287**, 251–2.

Godbole, S. S., Kaul, R., D'Souza, S. F. & Nadkarni, G. B. (1983). Immobilization of fumarase by entrapment of rat liver mitochondria in polyacrylamide gel using gamma rays. *Biotechnology and Bioengineering*, **25**, 217–24.

Gooday, G. W. (1978). The enzymology of hyphal growth. In *The Filamentous Fungi*, vol. 3, *Developmental Mycology*. ed. J. E. Smith & D. R. Berry, pp. 51–77. London: Edward Arnold.

Grizeau, D. & Navarro, J. M. (1986). Glycerol production by *Dunaliella tertiolecta* immobilized with Ca alginate beads. *Biotechnology Letters*, **8**, 261–4.

Grote, W., Lee, K. J. & Rogers, P. L. (1980). Continuous ethanol production by immobilised cells of *Zymomonas mobilis*. *Biotechnology Letters*, **2**, 481–6.

Gunning, B. E. S. & Robards, A. W. (1976). *Intercellular Communication in Plants: Studies on Plasmodesmata*. New York: Springer-Verlag.

Haggstrom, L. & Molin, N. (1980). Calcium alginate immobilized cells of *Clostridium acetobutylicum* for solvent production. *Biotechnology Letters*, **2**, 241–6.

Hahn-Hagerdal, B. (1981). Enzymes co-immobilized with micro-organisms for the microbial conversion of non-metabolizable substrate. *Acta Chemica Scandinavica*, *B*, **34**, 611–3.

Hahn-Hagerdal, B. (1983). Co-immobilization involving cells, organelles and enzymes. In *Immobilized Cells and Organelles*, vol. II, ed. B. Mattiasson, pp, 79–94. Boca Raton: CRC Press.

Hahn-Hagerdal, B., Andersson, E., Lopez-Leiva, M. & Mattiasson, B. (1981). Membrane biotechnology, co-immobilization, and aqueous two-phase systems; alternatives in bioconversion of cellulose. *Biotechnology and Bioengineering, Symposium no. 11*, ed. C. D. Scott, pp. 651–61. New York: John Wiley.

Hall, J. L., Flowers, T. S. & Roberts, R. M. (1981). *Plant Cell Structure and Metabolism*, 2nd edn. London: Longman.

Halwachs, W. (1979). The effect of pore diffusion on immobilized enzymes. *Process Biochemistry*, **14(6)**, 25–7.

Hancher, C. W. & Perona, J. J. (1982). Kinetic model for a fluidized-bed bioreactor for denitrification of wastewaters. *Biotechnology and Bioengineering, Symposium no. 12*, ed. E. L. Garden, Jr, pp. 317–26. New York: John Wiley.

Hannoun, B. J. M. & Stephanopoulos, G. (1986). Diffusion coefficients of glucose and ethanol in cell-free and cell-occupied calcium alginate membranes. *Biotechnology and Bioengineering*, **28**, 829–35.

Hartmeier, W. (1981). *Saccharomyces cerevisiae* coimmobilized with pepsin for wine making. In *Advances in Biotechnology*. (Proceedings of the 5th International Yeast Symposium held in London, Canada, 20–25 July 1980.) *Current Developments in Yeast Research*, ed. G. C. Stewart & I. Russell, pp. 105–10. Oxford: Pergamon Press.

Hartmeier, W. & Doppner, T. (1983). Preparation and properties of mycelium bound glucose oxidase co-immobilized with excess catalase. *Biotechnology Letters*, **5**, 743–8.

Hartmeier, W. & Heinrichs, A. (1986). Membrane enclosed alginate beads containing *Gluconobacter* cells and molecular dispersed catalase. *Biotechnology Letters*, **8**, 565-70.

Hawksworth, D. L. (1984). Papers in which fungi in culture are used: Circular letter from the Commonwealth Mycological Institute. *Biotechnology Letters*, **6**, 333–4.

Hikuma, M., Kubo, T., Yasuda, T., Karube, I. & Suzuki, S. (1979). Amperometric determination of acetic acid with immobilized *Trichosporon brassicae*. *Analytica Chemica Acta*, **109**, 33–8.

Hirtenstein, M. & Clark, J. (1983). Microcarrier-bound mammalian cells. In *Immobilized Cells and Organelles*, vol. I, ed. B. Mattiasson, pp. 57–88. Boca Raton: CRC Press.

Hirtenstein, M., Clark, J. & Lindgren, G. (1980). Microcarriers for animal cell culture: a brief review of theory and practice. *Developments in Biological Standardization*, **46**, 109–16.

Ho, C. S. (1986). An understanding of the forces in the adhesion of micro-organisms to surfaces. *Process Biochemistry*, **21**, 148–52.

Hollo, J., Toth, J., Tengerdy, R. P. & Johnson, J. E. (1979). Denitrification and removal of heavy metals from wastewater by immobilized micro-organisms. In *Immobilized Microbial Cells: ACS Symposium Series 106*, ed. K. Venkatsubramanian, pp. 73–86. Washington, DC: American Chemical Society.

Hopkinson, J. (1983). Hollow fibre cell culture applications in industry. In *Immobilized Cells and Organelles*, vol. I, ed. B. Mattiasson, pp. 89–99. Boca Raton: CRC Press.

Hsiao, H., Chiang, L., Yang, C., Chen, L. & Tsao, G. T. (1983). Preparation and performance of immobilized yeast cells in columns containing no inert carrier. *Biotechnology and Bioengineering*, **25**, 363–75.

Hu, J. (1986). Immobilization of cells containing glucose isomerase using a multifunctional cross-linking reagent. *Biotechnology Letters*, **8**, 127–30.

Huysman, P., Van Meenen, P., Van Assche, P. & Verstraete, W. (1983). Factors affecting the colonization of non-porous and porous packing materials in model upflow methane reactors. *Biotechnology Letters*, **5**, 643–8.

Inouye, M. (1980). *Bacterial Outer Membranes*. Chichester: John Wiley.

Irle, C., Piguet, P. F. & Vassalli, P. (1978). *In vitro* maturation of immature thymocytes into immunocompetent T cells in the absence of direct thymic influence. *Journal of Experimental Medicine*, **148**, 32–45.

Jian, H. R., Zhang, X., Lo, X. H., Zhou, S. P., Cai, X. W. & Tso, W. W. (1983). Porous kaolinite beads: A novel but economical solid support material for biomass in fluidised bed technology. *Biotechnology Letters*, **5**, 85–8.

Jirku, V., Macek, T., Vanek, T., Krumphanzl, V. & Kubanek, V. (1981). Continuous production of steroid glycoalkaloids by immobilized plant cells. *Biotechnology Letters*, **3**, 447–50.

Jirku, V., Turkova, J. & Krumphanzl, V. (1980). Immobilization of yeast cells with retention of cell division and extracellular production of macromolecules. *Biotechnology Letters*, **2**, 509–13.

Jirku, V., Turkova, J., Veruovic, B. & Kubanek, V. (1980). Immobilization of yeast cells on polyphenyleneoxide. *Biotechnology Letters*, **2**, 451–4.

Johansen, A. & Flink, J. M. (1986). A new principle for immobilized yeast reactors based on internal gelation of alginate. *Biotechnology Letters*, **8**, 121–6.

Jones, W. J., Guyot, J. & Wolfe, R. S. (1984). Methanogenesis from sucrose by defined immobilized consortia. *Applied and Environmental Microbiology*, **47**, 1–6.

Jones, A. & Veliky, I. A. (1981). Effect of medium constituents on the viability of immobilized plant cells. *Canadian Journal of Botany*, **59**, 2095–101.

Joshi, S. & Yamazaki, H. (1986). Cellulose acetate entrapment of *Escherichia coli* on cotton cloth for aspartate production. *Biotechnology Letters*, **8**, 277–82.

Juneja, L. R., Terasawa, M., Yamane, T. & Shimizu, S. (1986). Continuous ethanol fermentation using flocculent yeast entrapped in horizontal parallel flow bioreactor system. *Biotechnology Letters*, **8**, 431–6.

Karube, I., Kuriyama, S., Matsunaga, T. & Suzuki, S. (1980). Methane production from wastewater by immobilized methanogenic bacteria. *Biotechnology and Bioengineering*, **22**, 847–58.

Karube, I., Matsunaga, T., Nakahara, T., Suzuki, S. & Kada, T. (1981). Preliminary screening of mutagens with a microbial sensor. *Analytical Chemistry*, **53**, 1024–6.

Kasche, V. (1979). Intraparticle diffusion limitations. In *Characterization of Immobilized Biocatalysts*, ed. K. Buchholz, pp. 224–41. New York: Verlag Chemie.

Kasche, V. & Buchholz, K. (1979). Mass transfer influence on effectiveness. In *Characterization of Immobilized Biocatalysts*, ed. K. Buchholz, pp. 208–23. New York: Verlag Chemie.

Kastl, P. R., Baricos, W. H., Cohen, W. & Chambers, R. P. (1979). Hollow fiber entrapped microsomes as a liver assist device in drug overdose treatment. In *Immobilized Microbial Cells, ACS Symposium Series 106*, ed. K. Venkatsubramanian, pp. 239–47. Washington, DC: American Chemical Society.

Katinger, H. W. D. (1977). New fermenter configurations. In *Biotechnology and Fungal Differentiation*, FEMS Symposium 4, ed. J. Meyrath & J. D. Bu'Lock. New York: Academic Press.

Kaul, R., D'Souza, S. F. & Nadkarni, G. B. (1983). Bactericidal effect of hen egg white support: A step towards the use of self-sterilizing enzyme supports. *Biotechnology and Bioengineering*, **25**, 887–9.

Kawabata, Y. & Demain, A. L. (1979). Enzymatic synthesis of pantothenic acid by *Escherichia coli* cells. In *Immobilized Microbial Cells, ACS Symposium Series 106*, ed. K. Venkatsubramanian, pp. 133–7. Washington, DC: American Chemical Society.

Kennedy, J. F. (1979). Facile methods for the immobilization of microbial cells without disruption of their life processes. In *Immobilized Microbial Cells ACS Symposium Series 106*, ed. K. Venkatsubramanian, pp. 119–31. Washington, DC: American Chemical Society.

Kennedy, J. F., Humphreys, J. D., Barker, S. A. & Greenshields, R. N. (1980). Application of living immobilized cells to the acceleration of the continuous conversions of ethanol (wort) to acetic acid (vinegar) – hydrous titanium (IV) oxide-immobilized *Acetobacter* species. *Enzyme and Microbial Technology*, **2**, 209–16.

Kennedy, K. J. & Van Den Berg, L. (1982). Continuous vs slug loading of downflow stationary film reactors digestion piggery waste. *Biotechnology Letters*, **4**, 137–42.

Khachatourians, G. G., Brosseau, J. D. & Child, J. J. (1982). Thymidine phosphorylase activity of anucleate minicells of *Escherichia coli* immobilized in an agarose gel matrix. *Biotechnology Letters*, **4**, 735–40.

Kierstan, M. & Bucke, C. (1977). The immobilization of microbial cells, subcellular organelles and enzymes in calcium alginate gels. *Biotechnology and Bioengineering*, **19**, 387–97.

Kierstan, M., Darcy, G. & Reilly, J. (1982). Studies on the characteristics of alginate gels in relation to their use in separation and immobilization applications. *Biotechnology and Bioengineering*, **24**, 1507–17.

Kim, M. N., Ergan, F., Dhulster, P., Strat, P., Gellf, G. & Thomas, D. (1982). Steroid modification with immobilized mycelium of *Aspergillus phoenicis*. *Biotechnology Letters*, **4**, 233–8.

Klein, J. & Eng, H. (1979). Immobilization of microbial cells in epoxy carrier systems. *Biotechnology Letters*, **1**, 171–6.

Klein, J., Hackel, V. & Wagner, F. (1979). Phenol degradation by *Candida tropicalis* whole cells entrapped in polymeric ionic networks. In *Immobilized Microbial Cells, ACS Symposium Series 106*, ed. K. Venkatsubramanian, pp. 101–18. Washington, DC: American Chemical Society.

Klein, J., Hackel, U., Schara, P., Washausen, P., Wagner, F. & Martin, C. K. A. (1978). Polymer entrapment of microbial cells: preparation and reactivity of catalytic systems. In *Enzyme Engineering*, vol. 4, ed. G. B. Broun, G. Manecke & L. M. Wingard, pp. 339–41. New York: Plenum Press.

Klein, J., & Kluge, M. (1979). Packed bed flow resistance. In *Characterization of Immobilized Biocatalysts*, ed. K. Buchholz, pp. 285–91. New York: Verlag Chemie.

Klein, J. & Kluge, M. (1981). Immobilization of microbial cells in polyurethan matrices. *Biotechnology Letters*, **3**, 65–70.

Klein, J. & Kressdorf, B. (1982). Immobilization of living whole cells in an epoxy matrix. *Biotechnology Letters*, **4**, 357–80.

Klein, J. & Kressdorf, B. (1983). Improvement of productivity and efficiency in ethanol production with Ca-alginate immobilized *Zymomonas mobilis*. *Biotechnology Letters*, **5**, 497–502.

Klein, J. & Kressdorf, B. (1986). Rapid ethanol fermentation with immobilized *Zymomonas mobilis* in a three stage reactor system. *Biotechnology Letters*, **8**, 739–44.

Kloosterman, J. & Lilly, M. D. (1986). An airlift loop reactor for the transformation of steroids by immobilized cells. *Biotechnology Letters*, **7**, 25–30.

Kluge, M., Klein, J. & Wagner, F. (1982). Production of 6-aminopenicillanic acid by immobilized *Pleurotus ostreatus*. *Biotechnology Letters*, **4**, 293–6.

Kolot, F. B. (1981). Microbial carriers – strategy for selection. *Process Biochemistry*, **16**, 2–9, 30–46.

Krouwel, P. G., Groot, W. J. & Kossen, N. W. F. (1983). Continuous IBE fermentation by immobilized growing *Clostridium beijerinckii* cells in a stirred-tank fermentor. *Biotechnology and Bioengineering*, **25**, 281-99.

Krouwel, P. G., Harder, A. & Kossen, N. W. F. (1982). Tensile stress–strain measurements of materials used for immobilization. *Biotechnology Letters*, **4**, 103–8.

Krouwel, P. G., Van Der Laan, W. F. M. & Kossen, N. W. F. (1980). Continuous production of *n*-butanol and isopropanol by immobilized, growing *Clostridium butylicum* cells. *Biotechnology Letters*, **2**, 253–8.

Krouwel, P. G., Van Der Laan, W. F. M. & Kossen, N. W. F. (1981). A method for starting a fermentor with immobilized spores of the genus *Clostridium* aseptically. *Biotechnology Letters*, **3**, 153–4.

Krug, T. A. & Daugulis, A. J. (1983). Ethanol production using *Zymomonas mobilis* immobilized on an ion exchange resin. *Biotechnology Letters*, **5**, 159–64.

Kuhn, I. (1980). Alcoholic fermentation in an aqueous two-phase system. *Biotechnology and Bioengineering*, **22**, 2393–8.

Kumakura, M. & Kaetsu, I. (1983). Pre-coating of microbial cells by hydrophobic reagents on immobilization. *Biotechnology Letters*, **5**, 197–200.

Larkin, P. J. (1981). Plant protoplast agglutination and immobilization. In *The Phytochemistry of Cell Recognition and Cell Surface Interactions*, ed. F. A. Loewus & C. A. Ryan, pp. 135–60. New York: Plenum Press.

Larsson, P. O., Ohison, S. & Mosbach, K. (1979). Transformation of steroids by immobilized living microorganisms. In *Applied Biochemistry and Bioengineering*, vol. 2, *Enzyme Technology*, ed. L. B. Wingard, Jr, E. Katchalski-Katzir & L. Goldstein, pp. 291–301. New York: Academic Press.

Lee, K. J., Lefebvre, M., Tribe, D. E. & Rogers, P. L. (1980). High productivity ethanol fermentations with *Zymomonas mobilis* using continuous cell recycle. *Biotechnology Letters*, **2**, 487–92.

Lee, J. H., Skotnicki, M. L., & Rogers, P. L. (1982). Kinetic studies on a flocculent strain of *Zymomonas mobilis*. *Biotechnology Letters*, **4**, 615–20.

Lilly, M. D. (1983). Two-liquid phase biocatalytic reactors. *Philosophical Transactions of the Royal Society, London, Series B*, **300**, 391–8.

Lilly, M. D. (1986). Recommendations for nomenclature to describe the metabolic behaviour of immobilized cells. *Enzyme and Microbial Technology*, **8**, 315.

Lindsey, K. & Yeoman, M. M. (1983). Novel experimental systems for studying the production of secondary metabolites by plant tissue cultures. In *Plant Biotechnology*, ed. S. H. Mantell & H. Smith, pp. 39–66. Cambridge: Cambridge University Press.

Linko, Y. Y., Kautola, H., Uotila, S. & Linko, P. (1986). Alcoholic fermentation of D-xylose by immobilized *Pichia stipitis* yeast. *Biotechnology Letters*, **8**, 47–52.

Linko, Y. Y., Pohjola, L. & Linko, P. (1977). Entrapped glucose isomerase for high fructose syrup production. *Process Biochemistry*, **12(6)**, 14–16.

Linse, L. & Brodelius, P. (1983). Immobilization of *Daucus carota* protoplasts. In *6th International Protoplast Symposium*, ed. I. Potrykus *et al.*, pp. 260–1. Basel: Birkhauser.

Livernoche, D., Jurasek, L., Desrochers, M. & Veliky, I. A. (1981). Decolorization of a kraft mill effluent with fungal mycelium immobilized in calcium alginate gel. *Biotechnology Letters*, **3**, 701–6.

Livesey-Goldblatt, Tunley, T. H. & Nagy, I. F. (1977). Pilot-plant bacterial film oxidation (BAC Fox Process) of recycled acidified uranium plant ferrous sulphate leach solution. In *Conference on Bacterial Leaching, GBF Monograph Series 4*, ed. W. Schwartz, pp. 175–90. New York: Verlag Chemie.

Ljungdahl, L. G. (1983). Personal communication, quoted in Mattiasson, B. & Hahn-Hagerdol, B. (1983). Utilization of aqueous two-phase systems for generating soluble immobilized preparations of biocatalysts. In *Immobilized Cells and Organelles*, vol. 1, ed. B. Mattiasson, pp. 122–34. Boca Raton: CRC Press.

Luong, J. H. T. (1983). Determination of effectiveness factor for immobilized cells on solid supports. *European Journal of Applied Microbiology and Biotechnology*, **18**, 249–53.

Macaskie, L. E. & Dean, A. C. R. (1985a). Uranium accumulation by immobilized cells of a *Citrobacter* sp. *Biotechnology Letters*, **7**, 457–62.

Macaskie, L. E. & Dean, A. C. R. (1985b). Strontium accumulation by immobilized cells of a *Citrobacter* sp. *Biotechnology Letters*, **7**, 627–30.

McGhee, J. E., St. Julian, G., Detroy, R. W. & Bothast, R. J. (1982). Ethanol production by immobilized *Saccharomyces cerevisiae, Saccharomyces uvarum* and *Zymomonas mobilis*. *Biotechnology and Bioengineering*, **24**, 1155–63.

Maisterrena, B., Blum, J. & Coulet, P. R. (1986). A simple model analysis of product transfer through a non-impervious enzymic membrane. *Biotechnology Letters*, **8**, 305–10.

Marcipar, A., Cochet, N., Brackenridge, L. & Lebeault, J. M. (1979). Immobilization of yeasts on ceramic supports. *Biotechnology Letters*, **1**, 65–70.

Marek, M., Kas, J., Valentova, O., Demnerova, K. & Vodrazka, Z. (1986). Immobilization of cells via activated cell walls. *Biotechnology Letters*, **8**, 721–4.

Margaritis, A. & Bajpai, P. (1981). Repeated batch production of ethanol from jerusalem artichoke tubers using recycled immobilized cells of *Kluveromyces fragilis*. *Biotechnology Letters*, **3**, 679–82.

Margaritis, A., Bajpai, P. K. & Wallace, J. B. (1981). High ethanol productivity using small Ca-alginate beads of immobilized cells of *Zymomonas mobilis*. *Biotechnology Letters*, **3**, 613–18.

Margaritis, A. & Wallace, J. B. (1982). The use of immobilized cells of *Zymomonas mobilis* in a novel fluidized bioreactor to produce ethanol. In *Biotechnology and Bioengineering Symposium 12*, ed. E. L. Gaden, Jr, pp. 147–59. New York: John Wiley.

Marshak-Rothstein, A., Fink, P., Gridley, T., Raulet, D. H., Bevan, M. J. & Gester, M. L. (1979). Properties and applications of monoclonal antibodies directed against determinants of the Thy-1 locus. *Journal of Immunology*, **122**, 2491–7.

Marsot, P., Fournier, R. & Blais, C. (1981). Un nouveau procède de culture continue a dialyse pour le phytoplancton. *Biotechnology Letters*, **3**, 689–94.

Mason, J. R., Pirt, S. J. & Somerville, H. J. (1978). Benzene metabolism by bacterial cells immobilized in polyacrylamide gel. In *Enzyme Engineering*, vol. 4, ed. G. B. Broun, G. Manecke & L. M. Wingard. New York: Plenum Press.

Matsunaga, T., Karube, I. & Suzuki, S. (1978). Rapid determination of nicotinic acid by immobilized *Lactobacillus arabinosus*. *Analytica Chimica Acta*, **99**, 233–9.

Matsunaga, T., Karube, I., Teraoka, N. & Suzuki, S. (1981). Rapid determination of phenylalanine with immobilized *Leuconostoc mesenteroides* and a lactate electrode. *Analytica Chimica Acta*, **127**, 245–9.

Matteau, P. P. & Saddler, J. N. (1982a). Glucose production using immobilized mycelial-associated β-glucosidase of *Trichoderma* E58. *Biotechnology Letters*, **4**, 513–18.

Matteau, P. P. & Saddler, J. N. (1982b). Continuous glucose production using encapsulated mycelial-associated β-glucosidase from *Trichoderma* E58. *Biotechnology Letters*, **4**, 715–20.

Mattiasson, B. (1979). Application of immobilized whole cells in analysis. In

Immobilized Microbial Cells, ACS Symposium Series 106, ed. K. Venkatsubramanian, pp. 203–20. Washington, DC: American Chemical Society.

Mattiasson, B. (1983). *Immobilized Cells and Organelles*, vols 1 & 2. Boca Raton: CRC Press.

Mattiasson, B. & Hahn-Hagerdal, B. (1983). Utilization of aqueous two-phase systems for generating soluble immobilized preparations of biocatalysts. In *Immobilized Cells and Organelles*, vol I, ed. B. Mattiasson, pp. 121–34. Boca Raton: CRC Press.

Mattiasson, B., Larsson, P. O. & Mosbach, K. (1977). The microbe thermistor. *Nature*, **268**, 519–20.

Mattiasson, B., Ramstorp, M., Nilsson, I. & Hahn-Hagerdal, B. (1981). Comparison of the performance of a hollow-fibre microbe reactor with a reactor containing alginate entrapped cells. *Biotechnology Letters*, **3**, 561–6.

Matulovic, U., Rasch, D. & Wagner, F. (1986). New equipment for the scaled up production of small spherical biocatalysts. *Biotechnology Letters*, **8**, 485–90.

Mavituna, F. & Park, J. M. (1985). Growth of immobilized plant cells in reticulate polyurethane foam matrices. *Biotechnology Letters*, **7**, 637–40.

Mazia, D. (1974). The cell cycle. *Scientific American*, **230**(1), 54–64.

Messing, R. A. (1982). Immobilized microbes and a high rate continuous waste processor for the production of high rate gas and the reduction of pollutants. *Biotechnology and Bioengineering*, **24**, 1115–23.

Messing, R. A., Oppermann, R. A. & Kolot, F. B. (1979). Pore dimensions for accumulating biomass. In *Immobilized Microbial Cells, ACS Symposium Series 106*, ed. K. Venkatsubramanian, pp. 13–28. Washington, DC: American Chemical Society.

Meyerhoff, M. E. (1980). Preparation and response properties of selective bioelectrodes utilizing polymer membrane electrode-based ammonia-gas sensors. *Analytical Letters*, **13**(B15), 1345–58.

Michaux, M., Paquot, M., Baijot, B. & Thonart, P. (1982). Continuous fermentation: improvement of cell immobilization by zeta potential measurement. In *Biotechnology and Bioengineering Symposium 12*, ed. E. L. Gaden, Jr, pp. 475–84. New York: John Wiley.

Miyawaki, O., Nakamura, K. & Yano, T. (1978). Mass transfer and reaction with microcapsules containing enzyme and adsorbent. In *Enzyme Engineering*, vol. 3, ed. E. K. Pye & H. H. Weetall, pp. 79–84. New York: Plenum Press.

Miyawaki, O., Wingard, L. B., Jr, Brackin, J. S. & Silver, R. S. (1986). Formation of propylene oxide by *Nocardia corallina* immobilized in liquid paraffin. *Biotechnology and Bioengineering*, **28**, 343–8.

Mohamed, S. & Salleh, A. B. (1982). Physical properties of polyethyleneimine-alginate gels. *Biotechnology Letters*, **4**, 611–4.

Mohan, R. R. & Li, N. N. (1975). Nitrate and nitrate reduction by liquid membrane-encapsulated whole cells. *Biotechnology and Bioengineering*, **17**, 1137–56.

Morris, P. & Fowler, M. W. (1981). A new method for the production of fine plant cell suspension culture. *Plant Cell, Tissue and Organ Culture*, **1**, 15–24.

Morris, P., Scragg, A. H., Stafford, A. & Fowler, M. W. (1986). *Secondary Metabolism in Plant Cell Cultures*. Cambridge: Cambridge University Press.

Mosbach, K. (1976). *Methods in Enzymology*, vol. 44. New York: Academic Press.

Mosbach, K. & Mosbach, R. (1966). Entrapment of enzymes and micro-organisms in synthetic cross-linked polymers and their application in column techniques. *Acta Chemica Scandinavica*, **20**, 2807–10.

Moser, A. (1982). Bioreactors with thin-layer characteristics. *Biotechnology Letters*, **4**, 281–6.

Muallem, A., Bruce, D. & Hall, D. O. (1983). Photoproduction of H_2 and $NADPH_2$ by polyurethane-immobilized cyanobacteria. *Biotechnology Letters*, **5**, 365–8.

Murata, K., Tani, K., Kato, J. & Chibata, I. (1981). Glutathione production by immobilized *Saccharomyces cerevisiae* cells containing an ATP regeneration system. *European Journal of Applied Microbiology and Biotechnology*, **11**, 72–7.

Myerson, A. S. & Kline, P. (1983). The adsorption of *Thiobacillus ferrooxidans* on solid particles. *Biotechnology and Bioengineering*, **25**, 1669–76.

Navarro, J. M. & Durand, G. (1977). Modification of yeast metabolism by immobilization onto porous glass. *European Journal of Applied Microbiology*, **4**, 243–54.

Nazly, N. & Knowles, C. J. (1981). Cyanide degradation by immobilized fungi. *Biotechnology Letters*, **3**, 363–8.

Nicolaus, B., De Simone, A., Del Piano, L., Giardina, P. & Lama, L. (1986). Production of 2-keto-3-deoxygluconate by immobilized cells of *Sulfolobus solfataricus*. *Biotechnology Letters*, **8**, 497–500.

Nilsson, I., Ohlson, S., Haggstrom, L., Molin, N. & Mosbach, K. (1980). Denitrification of water using immobilized *Pseudomonas denitrificans* cells. *European Journal of Applied Microbiology*, **10**, 261–74.

Obuchi, K. & Maeda, H. (1985). Controlled flow reversal column reactor for microcapsules. *Biotechnology Letters*, **7**, 759–64.

Ochiai, H., Shibata, H., Fujishima, A. & Honda, K. (1979). Photocurrent by immobilized chloroplast film electrode. *Agricultural and Biological Chemistry*, **43**, 881–4.

Ochiai, H., Shibata, H., Sawa, Y. & Katoh, T. (1980). 'Living electrode' as a long-lived photoconverter for biophotolysis of water. *Proceedings of the National Academy of Science, USA*, **77**, 2442–4.

Oshima, T. (1978). Properties of heat stable enzymes of extreme thermophiles. In *Enzyme Engineering*, vol. 4, ed. G. B. Broun, G. Manecke & L. M. Wingard. New York: Plenum Press.

Papageorgiou, C. C. (1983). Immobilization of photosynthetically active intact chloroplasts in a cross-linked albumin matrix. *Biotechnology Letters*, **5**, 819–24.

Parascandola, P. & Scardi, V. (1981). Gelatin-entrapped whole-cell invertase. *Biotechnology Letters*, **3**, 369–74.

Paterson, S. L., Fane, A. G., Fell, C. J. D. & Rogers, P. I. R. (1986). Evidence of oxygen limitation in a hollow fibre reactor with immobilized *E. coli*. *Biotechnology Letters*, **8**, 561–6.

Paul, F. & Vignais, P. M. (1980). Photophosphorylation in bacterial chromatophores entrapped in alginate gel: improvement of physical and biochemical properties of gel beads with barium as gel-inducing agent. *Enzyme and Microbial Technology*, **2**, 281–7.

Pedersen, H. & Horvath, C. (1981). Open tubular heterogeneous enzyme reactors in continuous-flow analysis. In *Applied Biochemistry and Bioengineering*, vol. 3, *Analytical Application of Immobilized Enzymes and*

Cells, ed. L. B. Wingard, Jr, E. Katchalski-Katzir & L. Goldstein, pp. 1–96. New York: Academic Press.

Petersen, E. E. (1965). *Chemical Reaction Analysis*. Englewood Cliffs, New Jersey: Prentice-Hall.

Pharmacia (1980). *Cell Affinity Chromatography: Principles and Methods*. Uppsala: Pharmacia Fine Chemicals AB.

Pharmacia (1981). *Microcarrier Cell Culture: Principles and Methods*. Uppsala: Pharmacia Fine Chemicals AB.

Pickett, A. M., Topwala, H. J. & Bazin, M. J. (1979). A new method of industrial bioreactor operation: The Transient Operation Technique. *Process Biochemistry*, **14**, 10–16.

Pierrot, P., Fick, M. & Engasser, J. M. (1986). Continuous acetone–butanol fermentation with high productivity cell ultra-filtration and recycling. *Biotechnology Letters*, **8**, 253–6.

Pirt, S. J. (1975). *Principles of Microbe and Cell Cultivation*. Oxford: Blackwell Scientific Publishers.

Pollard, R. & Khosrovi, B. (1978). Reactor design for fermentation of fragile tissue cells. *Process Biochemistry*, **13**, 31–7.

Powell, M. S. & Slater, N. K. H. (1983). The deposition of bacterial cells from laminar flows onto solid surfaces. *Biotechnology and Bioengineering*, **25**, 891–900.

Prince, I. G. & Barford, J. P. (1982a). Continuous tower fermentation for power ethanol production. *Biotechnology Letters*, **4**, 263–8.

Prince, I. G. & Barford, J. P. (1982b). Induced flocculation of yeasts for use in the tower fermenter. *Biotechnology Letters*, **4**, 621–6.

Prince, I. G. & Barford, J. P. (1982c). Tower fermentation using *Zymomonas mobilis* for ethanol production. *Biotechnology Letters*, **4**, 525–30.

Puziss, M. & Heden, C. G. (1965). Toxin production by *Clostridium tetani* in biphasic liquid cultures. *Biotechnology and Bioengineering*, **7**, 355–66.

Pye, E. K. & Weetall, H. H. (1975). *Enzyme Engineering*, vol. 3. New York: Plenum Press.

Quayle, J. R. & Bull, A. T. (1982). *New Dimensions in Microbiology: Mixed Substrates, Mixed Cultures and Microbial Communities*. London: The Royal Society.

Ramirez, A. & Boudarel, M. J. (1983). Continuous production of ethanol on beet juice by a flocculent strain of *Saccharomyces cerevisiae*. *Biotechnology Letters*, **5**, 659–64.

Reed, R. H., Warr, S. R. C., Kerby, N. W. & Stewart, W. D. P. (1986). Osmotic shock-induced release of low molecular weight metabolites from free-living and immobilized cyanobacteria. *Enzyme and Microbial Technology*, **8**, 101–4.

Rehg, T., Dorger, C. & Chau, P. C. (1986). Application of an atomizer in producing small alginate gel beads for cell immobilization. *Biotechnology Letters*, **8**, 111–14.

Reuveny, S., Mizrahi, A., Kotler, M. & Freeman, A. (1983). Factors effecting cell attachment, spreading and growth on derivatized microcarriers. I. Establishment of working system and effect of the type of the amino-charged groups. *Biotechnology and Bioengineering*, **25**, 409–80.

Rhodes, M. J. C., Robins, R. J. & Payne, J. (1982). Progress in the immobilization of plant cells – the problems and potential of immobilised systems. Presented at the Plant Cell Culture Conference held on 7 December

1982 at the Sudbury Conference Centre, London. Organised by Oyez Scientific and Technical Services Ltd.

Richardson, J. J. & Zaki, W. N. (1954). Sedimentation and fluidization, part 1. *Transactions of the Institute for Chemical Engineers*, 32, 35–53.

Roberts, E. H. (1972). Cytological, genetical and metabolic changes associated with loss of viability. In *Viability of Seeds*, ed. E. H. Roberts, pp. 252–306. London: Chapman & Hall.

Rochefort, W. E., Rehg, T. & Chau, P. C. (1986). Trivalent cation stabilization of alginate gel for cell immobilization. *Biotechnology Letters*, 8, 115–20.

Roels, J. A. & Van Tilberg, R. (1979). Temperature dependence of the stability and the activity of immobilized glucose isomerase. In *Immobilized Microbial Cells, ACS Symposium Series 106*, ed. K. Venkatsubramanian, pp. 147–72. Washington, DC: American Chemical Society.

Rose, A. H. & Harrison, J. S. (1969). *The Yeasts*, vol. I, *Biology of the Yeasts*, New York: Academic Press.

Rosevear, A. (1982). *UK Patent Application* no. 2,083,825 and *UK Patent Application* no. 2,083,827.

Roy, T. B. V., Blanch, H. W. & Wilke, C. R. (1982). Lactic acid production by *Lactobacillus delbreuckii* in a hollow fibre fermenter. *Biotechnology Letters*, 4, 483–8.

Roy, T. B. V., Mandel, D. K., Dea, D. K., Blanch, H. W. & Wilke, C. R. (1983). The application of cell recycle to continuous fermentative lactic acid production. *Biotechnology Letters*, 5, 665–70.

Royer, G., Livernoche, D., Desrocher, M., Jurasek, L., Rouleau, D. & Mayer, R. C. (1983). Decolorization of Kraft mill effluent: Kinetics of a continuous process using immobilized *Coriolus versicolor*. *Biotechnology Letters*, 5, 321–6.

Rutter, P. R. & Leech, R. (1980). The deposition of *Streptococcus sanguis* NCTC 7868 from a flowing suspension. *Journal of General Microbiology*, 120, 301–7.

Sakimae, A. & Onishi, H. (1981). Preparation of immobilized enzymes of micro-organisms. U.S. Patent no. 4,276,381.

Salmon, J. M. (1984). Application of the technique of cellular permeabilization to the study of the enzymatic activities of *Saccharomyces cerevisiae* in continuous alcoholic fermentation. *Biotechnology Letters*, 6, 43–8.

Sarkar, J. M. & Mayaudon, J. (1983). Alanine synthesis by immobilized *Corynebacterium dismutans* cells. *Biotechnology Letters*, 5, 201–6.

Scheurich, P., Schnabl, H., Zimmermann, U. & Klein, J. (1980). Immobilization and mechanical support of individual protoplasts. *Biochimica et Biophysica Acta*, 598, 645–51.

Schubert, F., Wollenberger, U. & Scheller, F. (1983). Plant tissue-based amperometric tyrosine electrode. *Biotechnology Letters*, 5, 239–42.

Schulz, R., Krafft, H. & Lehmann, J. (1986). Experiences with a new type of microcarrier. *Biotechnology Letters*, 8, 557–60.

Scott, C. D., Hancher, C. W. & Shumate, S. E. (1978). A tapered fluidized bed as a bioreactor. In *Enzyme Engineering*, vol. 3, ed. E. K. Pye & H. H. Weetall, pp. 4–8. New York: Plenum Press.

Shillito, R. D., Paszkowski, J. & Potrykus, I. (1983). Culture in agarose improves protoplast plating and proliferation and permits division in otherwise unresponsive systems. In *6th International Protoplast Symposium*, ed. J. Potrykus *et al.*, pp. 266–7. Basel: Birkhauser.

Shimizu, S., Tani, Y. & Yamada, H. (1979). Synthesis of coenzyme A by immobilized bacterial cells. In *Immobilized Microbial Cells, ACS Symposium Series 106*, ed. K. Venkatsubramanian, pp. 87–100. Washington, DC: American Chemical Society.

Sidwell, J. S. & Rechnitz, G. A. (1985). 'Bananatrode' – an electrochemical biosensor for dopamine. *Biotechnology Letters*, **7**, 419–22.

SivaRaman, H., Rao, B. S., Pundle, A. V. & SivaRaman, C. (1982). Continuous ethanol production by yeast cells immobilized in open pore gelatin matrix. *Biotechnology Letters*, **4**, 359–64.

Smit, N. & Rechnitz, G. A. (1984). Leaf based biocatalytic membrane electrodes. *Biotechnology Letters*, **6**, 209–14.

Smith. J. E. & Berry, D. R. (1978). *The Filamentous Fungi*, vol. 3, *Developmental Mycology*. London: Edward Arnold.

Solyom, P. (1976). Kontinuerlig overvakning av akuttoxicitet i avloppsvatten. *Vatten*, **2**/76, 192.

Spier, R. E. (1980). Recent developments in the large scale cultivation of animal cells in mono-layers. *Advances in Biochemical Engineering*, **13**, 119–62.

Stenroos, S. L., Linko, Y. Y. & Linko, P. (1982). Production of L-lactic acid with immobilized *Lactobacillus delbrueckii*. *Biotechnology Letters*, **4**, 159–64.

Stephenson, J. P. & Murphy. (1980). Kinetics of biological fluidized bed waste-water denitrification. *Progress in Water Technology*, **12**, 159–71.

Stocklein, W. Eisgruber, A. & Schmidt, H. L. (1983). Conversion of L-phenylalanine to L-tyrosine by immobilized bacteria. *Biotechnology Letters*, **5**, 703–8.

Strandberg, G. W., Donaldson, T. L. & Arcuri, E. J. (1982). Continuous ethanol production by a flocculent strain of *Zymomonas mobilis*. *Biotechnology Letters*, **4**, 347–52.

Strobel, J. R., Jr, Ciavarelli, L. M., Starnes, R. L. & Lanzilotta, R. P. (1983). Biocatalytic synthesis of esters using dried *Rhizopus arrhizus* mycelium as a source of enzyme. Abstract 053 of the Annual American Society for Microbiology Meeting, p. 248.

Suidan, M. T., Cross, W. H. & Fong, M. (1980). Continuous bioregeneration of granular activated carbon during the anaerobic degradation of catechol. *Progress in Water Technology*, **12**, 203–14.

Sundaram, P. V. & Pye, E. K. (1978a). Recommendations for standardization of nomenclature in affinity chromatography. In *Enzyme Engineering*, vol. 3, ed. E. K. Pye & H. H. Weetall, pp. 553–6. New York: Plenum Press.

Sundaram, P. V. & Pye, E. K. (1978b). Matters arising from the nomenclature of immobilized enzymes. In *Enzyme Engineering*, vol. 3, ed. E. K. Pye & H. H. Weetall, pp. 557–61. New York: Plenum Press.

Suzuki, S. & Karube, I. (1978). Microbial electrode BOD sensor. In *Enzyme Engineering*, vol. 4, ed. G. B. Broun, G. Manecke & L. M. Wingard, pp. 329–33. New York: Plenum Press.

Suzuki, S. & Karube, I. (1979a). Microbial electrode sensors for cephalosporins and glucose. In *Immobilized Microbial Cells, ACS Symposium Series 106*, ed. K. Venkatsubramanian, pp. 221–36. Washington, DC: American Chemical Society.

Suzuki, S. & Karube, I. (1979b). Production of antibiotics and enzymes by immobilized whole cells. In *Immobilized Microbial Cells, ACS Symposium Series 106*, ed. K. Venkatsubramanian, pp. 59–72. Washington, DC: American Chemical Society.

Suzuki, S. & Karube, I. (1981). Bioelectrochemical sensors. In *Applied Biochemistry and Bioengineering*, vol. 3, ed. L. B. Wingard, Jr, E. Katchalski-Katzir & L. Goldstein, pp. 145–74. New York: Academic Press.

Tanaka, A. & Fukui, S. (1983). Immobilized organelles. In *Immobilized Cells and Organelles*, vol. 1, ed. B. Mattiasson, pp. 101–19. Boca Raton: CRC Press.

Tharakan, J. P. & Chau, P. C. (1986a). A radial flow hollow fiber bioreactor for the large-scale culture of mammalian cells. *Biotechnology and Bioengineering*, **28**, 329–42.

Tharakan, J. P. & Chau, P. C. (1986b). Operation and pressure distribution of immobilized cell hollow fiber bioreactors. *Biotechnology and Bioengineering*, **28**, 1064–71.

Thompson, I. M. & Forster, C. F. (1983). Nutrient conditions in relation to bacterial flocculation, extracellular polymer production and sludge settlement. *Biotechnology Letters*, **5**, 761–6.

Tramper, J., Van Der Plas, H. C., Van Der Kaaden, A., Muller, F. & Middlehoven, W. J. (1979). Xanthine oxidase activity of *Arthrobacter* X-4 cells immobilized in glutaraldehyde-crosslinked gelatin. *Biotechnology Letters*, **1**, 397–402.

Tso, W. W. & Fung, W. P. (1981). Bacterial organisms suitable for filamentous cell immobilization. *Biotechnology Letters*, **3**, 421–4.

Uribelarrea, J. L., Pacaud, S. & Goma, G. (1985). New method for measuring the cell water content by thermogravimetry. *Biotechnology Letters*, **7**, 75–80.

Van Altena, C. (1975). Cleaning non-cellulosic ultrafiltration membranes. *Process Biochemistry*, **10**, 26–9.

Vandamme, E. J. (1983). Peptide antibiotic production through immobilized biocatalyst technology. *Enzyme and Microbial Technology*, **5**, 403–16.

Van Den Berg, L. & Kennedy, K. J. (1981). Support materials for stationary fixed film reactors for high-rate methanogenic fermentation. *Biotechnology Letters*, **3**, 165–70.

Van Oss, C. J., Gillman, G. F. & Newmann, A. A. (1975). *Phagocytic Engulfment and Cell Adhesiveness*. New York: Dekker.

Van Wezel, A. L. (1967). Growth of cell strains and primary cells on microcarriers in homogeneous culture. *Nature*, **216**, 64–5.

Varani, J., Dame, M., Beals, T. F. & Wass, J. A. (1983). Growth of three established cell lines on glass microcarriers. *Biotechnology and Bioengineering*, **25**, 1359–72.

Venkatsubramanian, K., Karkare, S. B. & Vieth, W. R. (1983). Chemical Engineering Analysis of Immobilized Cell Systems. In *Applied Biochemistry and Bioengineering*, vol. 4, *Immobilized Microbial Cells*, ed. I. Chibata & L. B. Wingard, Jr, pp. 312–50. New York: Academic Press.

Vidoli, R., Yamazaki, H., Nasim, A. & Veliky, I. A. (1982). A novel procedure for the recovery of hybrid products from protoplast fusion. *Biotechnology Letters*, **4**, 781–4.

Vieth, W. R. & Venkatsubramanian, K. (1979). Immobilized microbial cells in complex biocatalysis. In *Immobilized Microbial Cells, ACS Symposium Series 106*, ed. K. Venkatsubramanian, pp. 1–11. Washington, DC: American Chemical Society.

Vijaikishore, P. & Karanth, N. G. (1986). Glycerol production by immobilized cells of *Pichia farinosa*. *Biotechnology Letters*, **8**, 257–60.

Vollbrecht, D. (1980). Oxygen deficiency and excretion of metabolites by strictly aerobic bacteria. *Biotechnology Letters*, **2**, 49–54.

Vorlop, K. D. & Klein, J. (1981). Formation of spherical chitosan biocatalysts by ionotropic gelation. *Biotechnology Letters*, **3**, 9–14.

Wada, M., Kato, J. & Chibata, I. (1980). Continuous production of ethanol using immobilized growing yeast cells. *European Journal of Applied Microbiology and Biotechnology*, **10**, 275–87.

Wagner, F. & Lang, S. (1979). *Microbial Characterization of Immobilized Biocatalysts*, ed. K. Buchholz, pp. 315–36. Dechema: Verlag Chemie.

Wagner, F. & Vogelmann, H. (1977). Cultivation of plant tissue cultures in bioreactors and formation of secondary metabolites. In *Plant Tissue Culture and its Biotechnological Application*, ed. W. Barz, E. Reinhard & M. H. Zenk, pp. 245–52. New York: Springer–Verlag.

Wang, H. Y. & Hettwer, D. J. (1982). Cell immobilization in K-carrageenan with tricalcium phosphate. *Biotechnology and Bioengineering*, **24**, 1827–38.

Wang, H. Y., Lee, S. S., Takach, Y. & Caethon, L. (1982). Maximizing microbial cell loading in immobilized cell systems. In *Biotechnology and Bioengineering Symposium 12*, ed. E. L. Gaden, Jr, pp. 139–46. New York: John Wiley.

Wanner, O. & Gujer, W. (1986). A multispecies biofilm model. *Biotechnology and Bioengineering*, **28**, 314–28.

Watanabe, Y., Ishiguro, M. & Nishidome, K. (1980). Nitrification kinetics in a rotating biological disk reactor. *Progress in Water Technology*, **12**, 233–51.

Weaver, J. C., Perley, C. R., Reames, E. M. & Cooney, C. L. (1980). Temporarily immobilized microorganisms: rapid measurements using a mass spectrometer. *Biotechnology Letters*, **2**, 133–7.

Weaver, J. C., Reames, F. M., DeAlleaume, L., Perley, C. R. & Cooney, C. L. (1978). Continuous measurements on immobilized cells by a mass filter. In *Enzyme Engineering*, vol. 4, ed. G. B. Broun, G. Manecke & L. M. Wingard, pp. 403–4. New York: Plenum Press.

Webb, C., Fukuda, H. & Atkinson, B. (1986). The production of cellulase in a spouted bed fermentor using cells immobilized in biomass support particles. *Biotechnology and Bioengineering*, **28**, 41–50.

Weeks, M. G., Munro, P. A. & Spedding, P. L. (1982). Semi-continuous ethanolic fermentation using a novel yeast settling and recycle technique. *Biotechnology Letters*, **4**, 85–90.

Weeks, M. G., Munro, P. A. & Spedding, P. L. (1983). New concepts for rapid yeast settling. 1. Flocculation with an inert powder. *Biotechnology and Bioengineering*, **25**, 687–97.

Weetall, H. H., Sharma, B. P. & Detar, C. C. (1981). Photometabolic production of hydrogen from organic substrates by free and immobilized mixed cultures of *Rhodospirillum rubrum* and *Klebsiella pneumoniae*. *Biotechnology and Bioengineering*, **23**, 605–14.

Welch, G. R. (1978). Enzyme sequences in the living cell. In *Enzyme Engineering*, vol. 4, ed. G. B. Broun, G. Manecke & L. M. Wingard, pp. 289–97. New York: Plenum Press.

Wheatley, M. A. & Phillips, C. R. (1983a). The influence of internal and external diffusional limitations on the observed kinetics of immobilized whole bacterial cells with cell-associated β-glucosidase activity. *Biotechnology Letters*, **5**, 79–84.

Wheatley, M. A. & Phillips, C. R. (1983b). Temperature effects during polymerization of polyacrylamide gels used for bacterial cell immobilization. *Biotechnology and Bioengineering*, **25**, 623–6.

Wikstrom, P., Szwajcer, E., Brodelius, P., Nilsson, K. & Mosbach, K. (1982). Formation of α-keto acids from amino acids using immobilized bacteria and algae. *Biotechnology Letters*, **4**, 153–8.

Willets, A. (1986). Diol production by *Aeromonas hydrophila*: comparison of different immobilization techniques. *Biotechnology Letters*, **8**, 437–40.

Wingard, L. B., Jr, Katchalski-Katzir, E. & Goldstein, L. (1981). *Applied Biochemistry and Bioengineering*, vol. 3, *Analytical Applications of Immobilized Enzymes and Cells. New York: Academic Press.*

Wisniewski, J., Winnicki, T. & Majewska, T. (1983). Continuous transformation of benzaldehyde to benzyl alcohol by *Rhodotorula mucilaginosa* immobilized in an ultrafiltration cell. *Biotechnology and Bioengineering*, **25**, 1441–52.

Wittler, R., Baumgartl, H., Lubbers, D. W. & Schugerl, K. (1986). Investigations of oxygen transfer into *Penicillium chrysogenum* pellets by microprobe measurements. *Biotechnology and Bioengineering*, **28**, 1024–36.

Wyllie, A. H. (1981). Cell death: A new classification separating apoptosis from necrosis. In *Cell Death in Biology and Pathology*, ed. I. D. Bowen & R. A. Lockshin, pp. 9–34. London: Chapman & Hall.

Young, J. C. & Dahab, M. F. (1982). Operational characteristics of anaerobic packed-bed reactors. In *Biotechnology and Bioengineering Symposium no. 12*, ed E. L. Gaden, Jr, pp. 303–16. New York: John Wiley.

Index